Crisis-ready Leadership

Crisis-ready Leadership

Building Resilient Organizations and Communities

Bob Campbell, PE

Registered Office
John Wiley & Sons, Inc., 111 River Street, Hoboken, NJ 07030, USA

Editorial Office
111 River Street, Hoboken, NJ 07030, USA

For details of our global editorial offices, customer services, and more information about Wiley products visit us at www.wiley.com.

Wiley also publishes its books in a variety of electronic formats and by print-on-demand. Some content that appears in standard print versions of this book may not be available in other formats.

A catalogue record for this book is available from the Library of Congress

Hardback ISBN: 9781119700234; ePub ISBN: 9781119700258; ePDF ISBN: 9781119700241

Cover Image: © Buena Vista Images/Getty Images
Cover Design: Wiley

Set in 9.5/12.5pt STIXTwoText by Integra Software Services Pvt. Ltd, Pondicherry, India

Contents

Preface *xi*
Acknowledgements *xiii*

Introduction *1*

Part One   The Situation *7*

1 **Emerging Threats and Hazards** *9*
1.1 Incident Management Systems *9*
1.2 Threats, Hazards, and Incidents *9*
1.3 Spectrum of Incidents *10*
1.3.1 Pre-incident Phase *10*
1.3.2 Incident Phase *11*
1.3.2.1 The Emergency *11*
1.3.2.2 The Crisis *12*
1.3.2.3 The Disaster *14*
1.3.2.4 The Catastrophe *14*
1.3.3 Post-incident *14*
1.4 Identifying Conventional Threats and Hazards *15*
1.5 Thinking Outside the Box *15*

2 **Operating Contexts – Command and Coordination** *23*
2.1 Public Safety Context *23*
2.2 Private Sector Context *25*
2.2.1 Stationary Hazards *26*
2.2.2 Mobile Hazards *27*
2.3 Combined Public–Private Operating Context *28*
2.3.1 El Dorado Chemical Company Emergency Case Study *28*
2.3.2 West Fertilizer Company Crisis Case Study *29*
2.3.3 Superstorm Sandy Disaster Case Study *29*
2.3.4 COVID-19 (Global Incident of Significance) Case Study *29*
2.4 Escalating Scale of Incidents Lead to Multijurisdictional Response *31*
2.5 Challenges of Multijurisdictional Response *31*
2.6 Large Scale Disaster, Multinational Context, and Challenges *31*

2.7   Multidisciplined Response   *32*
2.7.1   Credentialing and Certifications   *32*
2.7.2   HazMat Response Training and Certification   *32*
2.7.3   Consensus Standards for Operations   *33*
2.7.4   Methods of Triage, Decontamination, and Sampling   *33*
2.7.5   HazMat Detection   *33*
2.7.6   Public Communication   *33*
2.7.7   Technical Specialists   *33*
2.7.8   Recovery Contractors   *34*
2.8   Unified Command   *34*

3   The Challenge Ahead   *37*
3.1   Confronting the Evolving Threats, Hazards, and Vulnerabilities   *37*
3.1.1   Expanding Scope of Threats and Hazards   *38*
3.1.2   Increasing Severity of Threats and Hazards   *38*
3.1.3   Vulnerable Populations   *39*
3.2   Preparing to Overcome the Challenges   *40*
3.2.1   Planning   *40*
3.2.2   Funding Systems   *41*
3.2.3   Organizational Approach   *42*
3.2.4   Rethinking the Norms   *42*
3.2.5   Whole Community Preparedness   *43*
3.3   Creating Conditions for Success before the Crisis   *43*
3.4   Leading through a Crisis   *45*
3.4.1   Establish the Goal and Objectives   *45*
3.4.2   Communicate the Goal and Incident Objectives   *45*
3.4.3   Assign Resources   *46*
3.4.4   Coordinate with Stakeholders   *46*
3.4.5   Monitor Progress and Control Outcomes   *46*
3.4.6   Communicate Again   *47*

4   Risk-Based Planning   *49*
4.1   Hazard Identification Risk Assessment Methodology   *49*
4.2   Risk Calculation   *50*
4.2.1   Likelihood Rating   *52*
4.2.2   Consequence Rating   *52*
4.3   Vulnerability Assessment   *53*

         Part Two   Decision-Making   *57*

5   Situational Awareness   *59*
5.1   What Is Situational Awareness?   *59*
5.2   Situational Awareness Model   *60*
5.2.1   Perception   *60*

5.2.2   Comprehension   *61*

5.2.3   Projection   *63*

5.2.4   Intervention   *64*

5.3   Relationship between Situational Awareness and Decision-Making   *65*

5.3.1   Task and Environmental Factors   *66*

5.3.2   Individual Factors   *67*

5.4   Application of the Situational Awareness Model   *67*

6   Decision Theory   *71*

6.1   Critical Thinking   *71*

6.1.1   Perception Applied   *72*

6.1.2   Comprehension Applied   *73*

6.1.3   Projection Applied   *75*

6.1.4   Problem-Solving   *76*

6.1.5   Courses of Action   *77*

6.2   Decision-making   *78*

6.2.1   Resource Constraints   *79*

6.2.2   Time Constraints   *79*

6.2.2.1   Decision-making: No Time Constraints   *80*

6.2.2.2   Decision-making: Time Constrained   *81*

6.2.2.3   Decision-making: Making Good Decisions   *82*

6.2.3   Decision-Making Models   *83*

6.2.3.1   Data Quality Objective Model   *83*

6.2.3.2   Subjective, Empirical Model   *85*

6.2.4   Implementing the Decision   *86*

7   Application of FEMA's Community Lifelines   *93*

7.1   Terminology   *94*

7.2   Community Lifelines Construct   *95*

7.3   Benefits of the Community Lifelines Toolkit   *96*

7.4   Limitations with the Community Lifelines Toolkit   *97*

7.5   Application of Community Lifelines from a Local Leadership Perspective   *98*

7.5.1   Lifeline Assessment and Status   *98*

7.5.2   Establishing Incident Priorities   *99*

7.5.3   Operationalizing Lifelines by Organizing Response Activities around Lines of Effort   *100*

7.5.4   Establish Additional Logistics and Resource Requirements   *100*

7.5.5   Reassess Lifeline Conditions and Status   *101*

7.5.6   Stabilization and Recovery   *102*

7.6   An Organizational Perspective   *102*

7.6.1   Adapting Lifelines and Components   *103*

7.6.2   Detecting and Reporting Unstable Components   *103*

7.6.3   Stabilizing through Lines of Effort   *103*

7.7   Application Summary   *104*

Part Three   Adversity to Sound Judgment *107*

8   **Crisis Stress and Effect on Judgment** *109*
8.1   Definitions and Spectrum of Stress *109*
8.2   Psychological Effects *112*
8.3   Physiological Effects *113*
8.4   The Effects of Stress on Judgment *114*

9   **Overcoming Stress to Optimize Performance** *119*
9.1   Health and Wellness *119*
9.2   Planning *120*
9.3   Organization *123*
9.4   Equipment, Technology, and Decision-Support Tools *125*
9.5   Training and Exercises *126*
9.6   Expertise, Competency, and Proficiency *127*

Part Four   Crisis Leadership *131*

10   **Profiles in Crisis Leadership** *133*
10.1   Crisis Leader Profiles *133*
10.2   Brock Long: Leading FEMA through Transformation while Supporting Federal Response to Hundreds of Disasters during the Most Extensive and Costly Disaster Season in US History *134*
10.3   Lieutenant General (Ret) H.R. McMaster: Counterinsurgency against Al Qaeda in Iraq *139*
10.4   Frank Patterson: Incident Commander during the West Fertilizer Company Incident Response *142*
10.5   Derrick Vick, President of Freedom Industries: Leading through a Financial Crisis and a Return to Prosperity *145*
10.6   Chad Hawkins: COVID-19 Response Incident Management Team Incident Commander *149*
10.7   Major General (Ret) Dana Pittard: Leading the Campaign against ISIS in Iraq *152*

11   **Attributes of a Crisis Leader** *157*
11.1   Essential Character Traits *157*
11.1.1 Humility *158*
11.1.2 Integrity *158*
11.1.3 People Smarts *159*
11.1.4 Moral Courage Builds Confidence *160*
11.2   Knowledge *161*
11.2.1 Self-awareness *161*
11.2.2 The Plan *162*
11.2.3 Situational Awareness with Leading Indicators *163*
11.2.4 Multidisciplinary Knowledge *164*
11.3   Skills *165*
11.3.1 Communication *165*
11.3.2 Complex Problem-Solving and Design Thinking *166*

11.3.3   Trust Building   *167*
11.3.4   Decisiveness   *167*
11.4   Additional Advice   *167*
11.5   Developing Crisis Leadership Skills   *169*

Part Five   A Safe and Secure Tomorrow   *177*

12   **Preparedness**   *179*
12.1   Preparedness Cycle   *179*
12.1.1   Assessment   *180*
12.1.2   Planning   *181*
12.1.3   Organizing and Equipping   *181*
12.1.4   Training   *182*
12.1.5   Exercise/Testing   *183*
12.1.6   Improvement   *184*
12.2   Preparedness Cycle Implementation   *185*

13   **Building Resilience**   *189*
13.1   A Resilience Model   *190*
13.2   Resilience Indicators   *190*
13.3   Traditional Response and Recovery Model vs. Simultaneous Recovery   *192*
13.4   A Systematic Approach to Resilience   *193*
13.4.1   Define the System   *193*
13.4.2   Identify Local, Relevant Hazards and Assess the Risk   *193*
13.4.3   Identify the Components of the System That Are Impacted by the Hazards   *194*
13.4.4   Identify the Interdependent Components and the Extent of Impact   *196*
13.5   Aggregate the Impact on All Components Based on Risk of Each Hazard   *198*
13.6   Determine Vulnerabilities within the System   *199*
13.7   Develop Strategies to Strengthen Components to Enhance Resilience   *200*
13.7.1   Evaluate and Prioritize Resilience Measures That Increase System-wide Resilience   *200*
13.7.2   Implement Actions to Increase Resilience   *204*
13.8   Disruptions-Theory of Constraints   *205*
13.9   Whole Community Approach   *205*
13.10   Setting Conditions for Successful Outcomes before the Incident   *207*

14   **Navigating from Crisis to Recovery**   *211*
14.1   Summary of a Leader's Role in Crisis Leadership   *211*
14.2   Pre-incident   *212*
14.3   During the Crisis   *214*
14.4   Post-crisis   *215*
14.5   Developing a Plan of Action   *216*
14.5.1   Organizational Readiness   *216*
14.5.2   Leadership Readiness   *217*
14.6   Conclusion   *217*

Index   *219*

Preface

Exercising effective leadership under dynamic and adverse situations can be challenging and requires one's ability to exercise situational awareness, emotional intelligence, and effective communication skills to make good decisions. However, leading through a crisis requires enhanced leadership skills and character traits due to the speed with which the situation grows and evolves, the size and multitude of different response organizations with varied interests, and escalation of adverse consequences. These factors complicate decision-making due to increased uncertainty, stress, relational complexity, and competing interests.

Recent disasters such as the East Palestine hazardous materials rail incident, COVID-19 pandemic, and various global conflicts demonstrate the need for competent, crisis-ready leaders. Poor leadership and decision-making will often exacerbate a crisis into a disaster with far-reaching cascading effects. Organizations and communities that are unprepared tend to suffer more severe consequences that impact life safety, health, property, the environment, the economy, and reputation for longer periods than their resilient and prepared counterparts.

This text is your roadmap to maturing as a crisis-ready leader. The text compiles leading best practices, theories, and proven leadership experiences to codify a pathway for developing and cultivating crisis-ready leaders. It covers theories and principles of crisis leadership, decision-making, navigating stress, organizational management, and collaboration. But it also provides methodologies for how to prepare for a crisis, build resilient organizations and communities that accelerate recovery, and develop crisis leadership skills.

This book builds on my prior work, "Overcoming Obstacles to Integrated Response Operations Among Incongruent Responders" from the *Handbook of Emergency Response: A Human Factors and Systems Engineering Approach* (New York: CRC Press, 2014) and several courses that I recently developed for FEMA related to crisis leadership, decision-making, economic recovery, and resilience. Drawing from nine years of military leadership experience and 18 years as founder and CEO of Alliance Solutions Group, Inc., a consulting company that engages corporate, military, and public safety organizations in emergency management for catastrophic events, I have gained valuable insight and understanding related to crisis leadership. These insights are organized and shared throughout this book to advance the practice of crisis leadership.

As you read this book and work through the application exercises at the end of each chapter, I want to encourage you to develop a professional development plan for maturing and honing your skills as a crisis-ready leader. I also want to encourage you to apply what you learn by building resilience within your organization or community. Filled with examples, case studies and crisis leader profiles, this text will aid you with operationalizing theories and methods within your context.

The book is divided into five parts. The first part provides context to the risks faced by organizations and communities. Part 2 covers the relationship between situational awareness and decision theory. Part 3 examines how stress impacts judgment and how to overcome the effects of stress. Part 4 provides example profiles of crisis leaders and how they navigated their crises consistent with the theories and methods described in earlier chapters. Part 5 concludes with practical steps in creating resilience and developing a personalized plan for maturing as a crisis-ready leader.

Each chapter is structured for the professional learner and begins with learning objectives which can be achieved through both reading the text and completing the application exercises. Since building resilience is a common theme, keys to resilience and a summary of key points aid the learner in understanding the text and recalling the most important information. References and additional reading point the learner to external resources to go deeper into a topic. The application questions and exercises encourage the learner to reflect and apply what they learn but also engage the learner with application of the material in their context.

While the public sector has developed a mature body of knowledge and standards on many of these topics, this text is written with both public and private sector practitioners in mind. This text will benefit graduate students, practitioners, and organizational leaders who desire to be prepared for the next crisis. Finally, it is my hope that those who read this book will play an active role in creating resilience within their sphere of influence. Crisis-ready leadership begins long before a crisis occurs. It is cultivated through studying best practices, analyzing lessons learned, and practicing application of skills and sound judgment. But most importantly, a successful crisis leader sets the conditions for successful and equitable outcomes by developing certain character traits and organizational resilience before the crisis.

Yorktown, VA, February 2023   *Robert K Campbell*

Acknowledgements

I would like to express my gratitude to my loving wife, Dr. Amy Campbell for supporting me through this multiyear project. I would also like to thank those at Alliance Solutions Group, Inc. for much of the foundational work that went into research, needs analyses, and course development that ultimately illuminated the growing interest and need for this text. I am also grateful to my publisher John Wiley and Sons, Inc. for their patience, guidance, and encouragement along the way. A special thanks to Katie O'Malley for reviewing the text multiple times, formatting figures, and conducting sanity checks, and to the leaders who were willing to share their experience and perspectives with me; thank you for your leadership, service, and contributions to this book: Lieutenant General (ret) H.R. McMaster, Brock Long, Major General (ret) Dana Pittard, Frank Patterson, Chad Hawkins, and Derrick Vick. Finally, I wish to thank my family for their patience, encouragement, and sacrifices while I invested in this project. This was only possible with your support.

Introduction

The year 2020 will be remembered for decades and studied through multiple lenses. The beginning of the year was marked with a new hazard: SARS-CoV-2, that may be classified as a sub-hazard of infectious disease. This novel betacoronavirus caused one of the worst global pandemics in human history, spawning cascading psychological, social, economic, health, and political effects throughout the world. At the end of 2020, COVID-19 (the disease caused by SARS-CoV-2) had infected 83.8 M, was hospitalizing over 6,000 people per week [1], and had killed 1,824,590 people worldwide [2].

The countermeasures were also devastating causing schools to shut down and move to an online format, cancelation of funeral and church gatherings, and the global GDP decreased by 6.7% [3]. In June 2020, the United States Congressional Budget Office estimated that COVID-19 will cost the US economy $8 trillion over the next decade. Four million US workers were considered "long-term unemployed," or unemployed for more than 27 weeks. As the government took action to bolster businesses and those who were temporarily unemployed, payments to both businesses and workers were abused. In some cases, unqualified businesses accepted government loans which were forgiven under specific conditions while unemployment payments increased so significantly, that many workers chose not to return to work, creating a labor supply shortage in many industries which in turn impacted supply chains in other industries; this resulted in inflation and business closures. Globally, somewhere between 88 and 115 million people are expected to fall back into in extreme poverty as a result of the pandemic [4]. Isolation and mandates such as physical distancing and mask wearing increased anxiety and drug use in the United States while limited mental-health resources reached capacity causing delays in access to care [5]. The impacts of this crisis continue to have far-reaching and cascading effects on all aspects of life even two years after its inception.

Decision makers have been taxed beyond their capacity as they have had to manage the crisis in their organization, locality, state, and country. Meanwhile external influences from the media, politics, and the public challenge decision makers. Aside from the severe consequences, the uncertainty surrounding the transmissibility and virulence of this virus has complicated the leaders' ability to make decisions with a high degree of certainty. Unlike well-studied chemicals, diseases, and radioactive isotopes, little is known about this novel virus except what is revealed through limited historic and ongoing research. Long-term physiological effects from the illness are unknown. Pharmaceutical companies are investing in vaccines and treatments with positive short-term results and promising therapeutics. But we only know what we know. As the body of knowledge expands, decision makers continue to adopt emerging best practices and decision criteria to support improved decision-making. Our tolerance for ambiguity is low. We desire certainty, predictability, and the right decisions.

Although the exact origin of this virus is still unknown, we do know it was being studied in a virology lab in Wuhan, China. It rapidly spread from person to person throughout China, Europe, the

Crisis-ready Leadership: Building Resilient Organizations and Communities, First Edition. Bob Campbell, PE.
© 2023 John Wiley & Sons, Inc. Published 2023 by John Wiley & Sons, Inc.

Americas, and then on to Africa, despite drastic decisions to shut down and impose strict limitations on public spaces in the early stages of the virus. The uncertainty of the situation, coupled with the unknown nature and behavior of the virus, as well as the lack of evidence on the effectiveness of countermeasures, led many leaders to make decisions that were ultimately ineffective. New variants of the virus have further increased uncertainty regarding vaccine efficacy, infectious dose, and mortality.

Over the last year, informatics and clarity have improved the methods used to analyze and support decision-making. As the pandemic becomes endemic, leaders will need to adapt further to the unpredictable and geographically dispersed waves of the disease that will require decisions based on time-phased, geographical variances in conditions. The growing matrix of control measures in vaccines, therapeutics, ventilation systems, and masking with widely varying efficacies will likely result in non-standard, novel approaches to risk management. Ultimately, leaders will need to lead with influence, vision, clarity, and decisiveness. History will judge whether the decisions were draconian or wise.

The vision for this book began long before the pandemic but this ongoing crisis which has spawned subsequent crises illustrates the growing need for crisis-ready leaders in all organizations and government offices. My journey in uncovering and ultimately addressing the need to train crisis-ready leaders began on a business trip in 2015. During a meeting with the Chief Resiliency Officer of a major international airline, she asked me whether my company, Alliance Solutions Group, Inc., could provide crisis management training and exercises. She went on to describe several crises that they had encountered and the unique risks of the airline industry operating in over a hundred countries around the world. She shared insight into how her company conducted exercises and explored the "what if" scenario based on how Malaysian airlines handled the disappearance of Flight MH370 on March 8, 2014. The need to train leaders in crisis management and decision-making was clear amidst the time-sensitive challenges – addressing multiple stakeholders, media, and governmental influences, timely and effective decision-making, risk management, and many more.

Over the last 17 years, my company has conducted thousands of training courses and exercises with military, first responders, hospitals, and Fortune 50 businesses throughout 49 states and 17 countries. This breadth of experience with diverse perspectives and varying levels of maturity has enabled me to observe and capture the challenges that leaders face during a crisis as well as the best practices.

In 2015, we were selected by FEMA as a sub-awardee on a training grant to address the root causes and contributing factors to several incidents of national significance. These included a growing number of crude-by-rail incidents and the West Fertilizer Company explosion in West, Texas. As I researched these incidents through discussions with stakeholders, regulators, and investigators at the Chemical Safety Board, it became clear that while first responders can leverage their experience and knowledge of incident command, there was a lack of training programs for first responders that developed leaders in crisis management. While knowledge was emphasized and measured in most training programs, and application of skills was emphasized in technician-level training, we uncovered that most courses did not achieve higher learning objectives such as analysis, synthesis, and evaluation. These application and judgment-related objectives are necessary in developing crisis leaders and decision makers. FEMA offered one course for elected officials on crisis leadership; this course relied on several case studies that occurred prior to Hurricane Katrina in 2005. So much has changed since 2005: FEMA was wrapped into the Department of Homeland Security, the National Response Framework was developed, the National Incident Management System matured to include the standardization of the Incident Command System, and so much more. So, we developed MGT-457, On-scene Crisis Leadership and Decision-Making for Hazmat Incidents. This course was delivered over 40 times during two years in 25 different states. One major city requested the course three times and maximized attendance!

We have also developed and delivered several other versions of this course for health-care leaders, business leaders, economic development organizations, local government leaders, and emergency management organizations.

Over the last 25 years, I have had several opportunities to study and apply leadership. From 1997 to 2005, I served as an officer in the US Air Force where I participated in various leadership programs and professional education that emphasized leadership development. After completing three assignments and achieving the rank of Major, I started a new career and launched Alliance Solutions Group, Inc. Over the last 17 years, I have continued my professional education in leadership with formal courses and participation in an executive peer advisory group. During the last 10 years, I have had the joy of serving on several Boards for non-profit organizations including a K4-12 private school. In 2014, I launched an international business to focus on the unique aspects of emergency management and response in the international context. It is through these experiences, that I have learned that while leadership knowledge and skills are a necessary foundation for crisis leadership, the crisis-ready leader requires additional knowledge, skills, and experience in order to succeed under the pressures that they will encounter. As I have also trained thousands of military personnel, first responders, and companies in emergency and crisis management and response, I have had the opportunity to uncover the keys to successful crisis leadership. The need for this book was validated through several of the experiences that I described above and throughout my career. My hope is to share what I have learned throughout the pages of this book in order to prepare crisis-ready leaders for the challenges that they will face.

How to Get the Most Out of This Book

The content of this book is organized topically so that it can be used in both educational and professional development contexts. I recommend reading the book in topic and chapter order as there are some foundational definitions and contexts provided in the earlier chapters that are built upon in later chapters. For example, the Profiles in Crisis Leadership unveils how each leader within their context addressed the topics presented earlier (e.g., maintaining situational awareness, decision-making, stress recognition and management, organizational challenges, external influences, etc.). Throughout the book, we will explore case studies from many different types of crises, including the COVID-19 global pandemic. This will help contextualize the topics with real-world examples. The objective is that our audience can quickly learn and apply the lessons underlying each topic. With that in mind, I have purposefully organized each chapter according to the following sections.

Learning Objectives

The learning objectives are designed to provide instructional expectations on what you should be able to achieve if you complete the reading and application sections of the chapter. While knowledge will come from reading the material, I am also providing opportunities for you to synthesize your experience (and the experiences of your cohort if you are working through this in a class or group study) with the content of the book. We learn from others and their experiences, so I recommend reading this book with others in order to get the most out of each chapter and application.

Main Body

Each chapter provides background theory, relevant content, and information on the topic. Where applicable, I have included case studies and examples, definitions, instruction on how to apply the information, best practices, and challenges facing professionals. While most chapters follow a similar framework, some chapters are organized according to the topic. A process will be explained in sequential steps. Profiles in Crisis Leadership presents each profile, and the content follows closely with the flow of my interview, the book topics, and my interpretation of their responses. While most questions are reserved for the end of the chapter, I did not want to miss an opportunity to interject a thought-provoking question in the middle of the text.

Summary of Key Points

A summary of key points is provided to aid the reader in recalling the most important information. These may also serve as discussion points related to sub-topics within the chapter.

Keys to Resilience

Since decision-making begins prior to the crisis, we will also explore how the operating environment impacts decision-making and vice versa with "keys to resilience." This section of each chapter highlights actions and initiatives that could enhance resilience for your community or organization. As you read and reflect on each topic and case study, consider what resiliency measures could have positively changed the outcomes. Building resilience is crucial to mitigating the impact of a crisis.

Application

To get the most out of this book, there are several application-based discussion points that may be helpful to further your learning on the topic. These are organized in a way that will align with increasingly robust learning objectives: discussion questions to enhance application skills, open-ended questions that require synthesis with outside information and experience, and exercises that require evaluation, formulation, and judgment. Not every topic lends itself to all of these types of discussion questions, but where applicable, I am providing the opportunity to go deeper. Exploring these application points with a cohort will further add to the value of these points through learning from the experience of others.

Further Reading

While I have included footnotes throughout the book, I also want to highlight several sources to explore for deeper knowledge on each topic. This enables me to focus on unique content while referring readers to supplemental works that also address the topic. Some chapters have a large body of knowledge related to the topic, while other chapters present novel concepts and may have fewer external sources.

It is my hope that this book not only strengthens your knowledge of crisis leadership and decision-making but also equips you to discern and comprehend reliable information, navigate uncertain and complex situations with confidence and effectiveness, and manage a crisis successfully with optimal outcomes. We live in an imperfect, complex, and interdependent world where there is rarely one simple right answer, but rather opportunities to formulate courses of action that maximize good outcomes for most people, while minimizing adverse effects. May this book prepare you to lead your organization or community through the next crisis and instill a commitment to investing in resilience.

References

1 U.S. Centers for Disease Control (2020). COVID-NET: a weekly summary of U.S. COVID-19 hospitalization data. https://gis.cdc.gov/grasp/covidnet/COVID19_5.html (accessed October 2, 2021).

2 The American Journal of Managed Care (2021). A timeline of COVID-19 developments in 2020. https://www.ajmc.com/view/a-timeline-of-covid19-developments-in-2020 (accessed October 2, 2021).

3 Statista (2021). Share of Gross Domestic Product (GDP) lost as a result of the coronavirus pandemic (COVID-19) in 2020, by economy. https://www.statista.com/statistics/1240594/gdp-loss-covid-19-economy (accessed October 2, 2021).

4 Atlantic Council (2020). The global economy in 2020, by the numbers. https://www.atlanticcouncil.org/blogs/new-atlanticist/the-global-economy-in-2020-by-the-numbers (accessed October 2, 2021).

5 Crane, L.D., Decker, R.A., Flaaen, A., et al. (2021). Business exit during the COVID-19 pandemic: non-traditional measures in historical context. Finance and Economics Discussion Series 2020-089r1. Washington: Board of Governors of the Federal Reserve System. doi: 10.17016/FEDS.2020.089r1.

Part One

The Situation

A current assessment of our situation includes a review of internal and external factors that could influence a crisis. We will begin with a survey of existing and emerging threats and hazards. These are external factors that could cause a crisis or affect the outcome of an incident. In Chapter 2, we will explore the internal factors such as our organization, its structure, and the liaison with external parties that we will interface with during a crisis. In Chapter 3, we will explore the challenges ahead that we need to address today. Finally, in chapter 4, we will ground our understanding of risk with a generally accepted process for conducting risk assessments so that we properly prioritize risks when planning, allocating resource, prioritizing preparedness initiatives, and making decisions.

1

Emerging Threats and Hazards

Learning Objectives

At the end of this chapter, you will be able to:
- Identify various types of incidents.
- Understand the resources available for responding to threats and hazards.
- Identify different levels of government responses to incidents.
- Identify key decisions made in response to various incidents.
- Describe elements that would make a community more resilient.

1.1 Incident Management Systems

The scope and scale of an incident demands a scalable systems-approach to resolving, managing, and recovering from the incident. This is often referred to as an Incident Management System. In the US, the Federal Emergency Management Agency (FEMA) has defined the components of the *National Incident Management System* as the following: National Qualification System, Resource Typing, Inventorying, Mutual Aid, Incident Command System, and Emergency Operations Centers. This standard approach ensures communities at the local, state, and federal levels of government can interact, communicate, share resources, rely on personnel and equipment from other jurisdictions as they coordinate resources to save lives, stabilize the incident, and protect property and the environment. Regardless of the system used, public and private sector organizations need to be able to recognize and respond appropriately when an incident occurs. The type, scale, and complexity of an incident will determine the type and number of responders.

1.2 Threats, Hazards, and Incidents

Incidents result from threats or hazards. A **threat** is a "natural, technological, or human-caused occurrence, individual, entity, or action that has or indicates the potential to harm life, information, operations, the environment, and/or property." As we study the nature and risk of threats, we must examine their consequences such as short- and long-term, secondary, and tertiary effects. These may include economic, brand or reputation of a company or jurisdiction, assets, systems, infrastructure, governmental capacities, safety, and security impacts. A **hazard** is "something that is

Crisis-ready Leadership: Building Resilient Organizations and Communities, First Edition. Bob Campbell, PE.
© 2023 John Wiley & Sons, Inc. Published 2023 by John Wiley & Sons, Inc.

potentially dangerous or harmful, often the root cause of an unwanted outcome." While hazards exist throughout our communities, they are often controlled or limited in a way that yields a benefit that outweighs the risk of an adverse outcome. For example, most people consider electricity and natural gas a necessity, an essential service provided by the utility company that creates some tangible benefit which outweighs the risk of an explosion or electrocution.

An **incident** is "an occurrence, natural or human-caused, that requires a response to protect life or property. Incidents can, for example, include major disasters, emergencies, terrorist attacks, terrorist threats, civil unrest, wildland and urban fires, floods, hazardous materials spills, nuclear accidents, aircraft accidents, earthquakes, hurricanes, tornadoes, tropical storms, tsunamis, war-related disasters, public health and medical emergencies, and other occurrences requiring an emergency response." So, threats and hazards can become incidents. These terms establish a basis for this chapter on emerging threats and hazards which provide a context for all-hazards incidents.

1.3   Spectrum of Incidents

The Spectrum of Incidents defines the scope and scale of an incident in order to categorize it into proper terms. As defined above, an incident is an "occurrence" – i.e., something that happens. This implies that there is a pre-incident phase and a post-incident phase. Figure 1.1 illustrates the progression of an incident throughout the spectrum of different types of incidents based on escalation. A prepared and resilient community with sound crisis leadership will be impacted less than a community marked by a lack of preparedness and poor crisis leadership decision-making.

1.3.1   Pre-incident Phase

During the pre-incident phase, communities have an opportunity to prepare for the incident. Leaders in government, business, academia, and other organizations are faced with a decision whether or not to prepare. This decision may not be a conscious decision, but it still takes place.

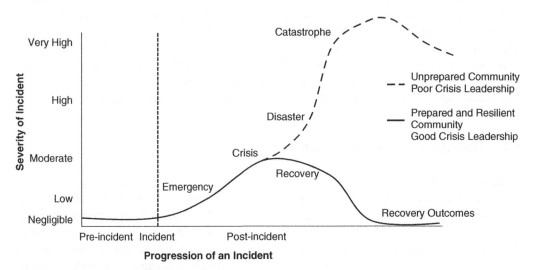

Figure 1.1   Spectrum of incidents.

This decision often involves known and unknown risks to the organization and community. Known risks can be managed whereas unknown risks are typically overlooked. After all, we cannot prepare for everything – or can we? Risk management can be a complex science that engages actuaries and futurists, or it could be a simple process of evaluating risks and determining the extent to which the organization is willing to expend resources and maintain a productive balance of activities. Pre-incident planning may involve activities to prevent an incident, mitigate the effects of an incident, or build resilience into the organization so that it rebounds from an incident. These pre-incident decisions can have a profound effect on establishing the conditions under which the organization will confront an incident.

1.3.2 Incident Phase

The incident phase can begin abruptly with an explosion, with some warning such as a tornado, or with much warning as an approaching winter storm. The scale of the incident can be small, such as a container of spilled oil, or something as large as a hurricane. The scope of the incident can be contained and controlled, such as the oil spill with absorbent pads, or uncontained and uncontrolled, such as an oil well leaking at the bottom of the ocean. The scope and scale of the incident will drive an initial decision on whether it can be handled internally or externally. The decision-maker will be taking into account their capability to handle the situation. Influencing factors may include trained response personnel, equipment, operational controls, psychological pressures such as concern for who will be held accountable, responsibility and decision-making authority, organizational policies and communications procedures, and safety culture.

1.3.2.1 The Emergency

An **emergency** is "any incident, whether natural, technological, or human-caused, that requires responsive action to protect life or property." For relatively small incidents, this typically involves a call to a 911 call center for police, fire, or EMS support. If the incident can be handled safely within the boundaries of private property (e.g., a chemical facility, gas station, etc.), then it has not likely met the threshold of an emergency. Some key criteria that facility operators may use to qualify an incident as an emergency may include:

- Is it uncontained?
- Is it uncontrolled?
- Is it unstable or escalating?
- Do we lack the resources to respond and recover?
- Does this incident pose a high risk to health, life, or property damage?
- Are there injuries, deaths, or property damage that threatens escalation of the incident?
- Is it a reportable incident?

An affirmative response to any of these questions may qualify the incident as an emergency. Specific knowledge and skills are needed to answer these questions accurately. Accurate knowledge requires a high degree of situational awareness of the incident, personnel status, capabilities, policies, and legal requirements. The situation also demands skilled decision-makers and responders to resolve the incident. These skills are built with clear policies, procedures, training, exercises, and communication systems. Decision-makers need to synthesize information from multiple sources, assess the risks, and evaluate the situation so that good judgment and decisions are made.

"In case of an emergency, dial 911" is frequently posted on signs as a reminder that this is the first action to take when an emergency is identified. After calling 911, the operator will usually ask,

"What is the emergency?" At this point, the caller is committed to describing the emergency and providing important information to the operator such as address, resources needed, status of those involved in the incident, and whether the incident involves a fire, chemical, or active threat.

Upon arrival of emergency services, the organization will need to quickly and efficiently provide information needed to ensure external responders obtain a high degree of situational awareness so that they can make informed and good decisions. In some cases, facilities have existing capabilities to engage in the response. However, in most cases, public safety responders arrive on scene and engage in response without support from private facility operators or owners. When there are multiple organizations or agencies involved in the response, there is a need to unify efforts and coordinate response activities. This is typically addressed in a unified command structure.

Most emergencies are resolved locally or regionally with public safety responders within a few hours or a single operational period. However, emergencies can escalate requiring additional resources. Train derailments involving hazardous materials are long-duration incidents that can extend over several days. Rural communities with limited public safety resources may be overwhelmed with the multitude of governmental agencies and related organizations that ultimately arrive to provide support. When an escalating incident is unaddressed, the escalation can reach a critical point at which the emergency transpires into a disaster. This relevant period of decision-making may be referred to as a crisis.

1.3.2.2 The Crisis

Crisis: "a hard and complicated situation ... or a turning point – a decisive crucial time/event, or a time of great danger or trouble with the possibilities of both good and bad outcomes" [1].

Some incidents begin as emergencies and then escalate into a crisis or disaster. Other incidents begin as a disaster with the potential of spawning other disasters if it is not managed properly. While other incidents begin as a crisis and require adept decision-making to de-escalate and avert a disaster.

An emergency that escalates into a crisis creates a stressful environment for leaders and decision-makers as the timeframe for making a decision and taking action is unknown. Since decision-making can either lead to incident recovery or disaster response, it is important to create the conditions for success in the pre-incident phase and make good enough decisions that manage the crisis to the point where the incident will de-escalate. Like a game of chess, aggressive and rapid decisions can lead to positive outcomes when well-planned and the player has a high degree of situational awareness. But when the next move is not thoroughly examined for adverse consequences, it could have disastrous results. When decision-makers know how much time they have to make a decision, they can manage the clock, but when this is unknown, they are operating with a higher degree of uncertainty.

Farazmand defines a **crisis** as follows: "Crises involve events and processes that carry severe threat, uncertainty, an unknown outcome, and urgency ... Most crises have trigger points so critical as to leave historical marks on nations, groups, and individual lives. Crises are historical points of reference, distinguishing between the past and the present. ... Crises come in a variety of forms, such as terrorism (New York World Trade Center and Oklahoma bombings), natural disasters (Hurricanes Hugo and Andrew in Florida, the Holland and Bangladesh flood disasters), nuclear plant accidents (Three-Mile Island and Chernobyl), riots (Los Angeles riot and the Paris riot of 1968, or periodic prison riots), business crises, and organizational crises facing life-or-death situations in a time of rapid environmental change. ... Crises consist of a 'short chain of events that destroy or drastically weaken' a condition of equilibrium and the effectiveness of a system or regime within a period of days, weeks, or hours rather than years. ... Surprises characterize the dynamics of crisis situations. ... Some crises are processes of events leading to a level of criticality or degree of intensity generally out of control. Crises often have past origins and diagnosing their original sources can help to understand and manage a particular crisis or lead it to alternative state of condition." [2]

The Cuban Missile crisis is accurately described as a crisis. This did not begin as a traditional emergency, but rather a situation that could have led to disastrous consequences if not confronted in a timely manner. Unlike most natural or technological hazards, this incident involved opposing, thinking, calculating opponents who were making moves and counter moves. This dynamic nature of the threat added more complexity to the decision-making process.

The COVID-19 pandemic quickly escalated to global disaster. But unlike a single disaster that passed, the ongoing and evolving nature of the threat and consequences qualifies the pandemic as a crisis. The right decisions could lead to recovery whereas the wrong decision could lead to a disaster. The complex, interdependent nature of this crisis with other facets of life such as the economy, employment, liberties, government services and institutions, private operations, etc. make this an even more challenging crisis to manage.

The key characteristics of a crisis are uncertainty, technical complexity, high severity, potential for large scale engagement, multifaceted, secondary effects, and urgency. While this book will explore crisis management in more detail, it is important to summarize some of the key ingredients for successful crisis management before exploring the disaster phase. Table 1.1 summarizes strategies for managing each of the key characteristics of a crisis.

Crisis Management: "Key to crisis management is an accurate and timely diagnosis of the criticality of the problems and the dynamics of events that ensue. This requires knowledge, skills, courageous leadership full of risk-taking ability, and vigilance. Successful crisis management also requires motivation, a sense of urgency, commitment, and creative thinking with a long-term strategic vision. In managing crises, established organizational norms, culture, rules and procedures become major obstacles: administrators and bureaucrats tend to protect themselves by playing a bureaucratic game and hiding behind organizational and legal shelters. A sense of urgency gives way to inertia and organizational sheltering and self-protection by managers and staff alike. ...

Table 1.1 Crisis characteristics and crisis management strategies.

Key Characteristics	Crisis Management Strategies
Uncertainty	Improve situational awareness with proven systems, experts, communication, common operating picture, analysis, forecasting
Technical complexity	Engage and trust experts for input, development of courses of action; zoom out and consider the big picture; employ strategic plans/thinking
High severity	Model/forecast courses of action to compare outcomes; account for all outcomes to include secondary effects
Potential for large scale engagement	Establish relationships with parties that are impacted and those that can support the response; consider temporal and geographic expansiveness
Multi-faceted	Integrate and engage multiple subject matter experts at the right level of decision-making, operations, and support to manage relevant elements of the incident
Secondary effects	Account for these in models and all courses of action
Urgency	Be decisive; continuously consider the risk of doing nothing with alternatives

Successful crisis management requires: (1) sensing the urgency of the matter; (2) thinking creatively and strategically to solving the crisis; (3) taking bold actions and acting courageously and sincerely; (4) breaking away from the self-protective organizational culture by taking risks and actions that may produce optimum solutions in which there would be no significant losers; and (5) maintaining a continuous presence in the rapidly changing situation with unfolding dramatic events." [2]

1.3.2.3 The Disaster

Like the crisis, the disaster can result from an escalating emergency or the wrong decisions during a crisis, or an occurrence so severe that the incident presents as a disaster. The West Fertilizer Company explosion at an Ammonium Nitrate processing facility in West, TX was an emergency that escalated to a disaster. After 20 minutes of attempting to fight the fire, the facility exploded killing 12 first responders and 3 members of the public, destroying 150 buildings, hospitalizing over 200 people, and causing $150M in damages. Likewise, the explosion at the port in Beirut, Lebanon was an immediate disaster which killed over 200 people, hospitalized 7,500 and displaced 300,000 people. The events that followed both incidents resulted in ongoing crises as concerns over the cause (i.e., potential for terrorism), heightened concerns during the incident management.

A **disaster** is "an occurrence of a natural catastrophe, technological accident, or human-caused event that has resulted in severe property damage, deaths, and/or multiple injuries."

Disasters require mobilization of national and international resources to respond and recover and often take extended time periods to fully recover and restore communities to their pre-incident status. Disasters are particularly deleterious to impoverished populations as they usually live in vulnerable locations, lack the financial and social resources to sustain themselves outside of their home, and require outside assistance for basic needs. Additionally, there are several other socially vulnerable populations which require further consideration when building a culture of preparedness throughout the community. Factors to consider include: non-English speakers, literacy, age (i.e., elderly and children), accessibility to transportation, and physical and mental disabilities.

1.3.2.4 The Catastrophe

While not defined in the Stafford Act, there are a couple of terms that signify an elevated disaster such as a catastrophe and incident of global significance. A **catastrophe** is an "event that results in large numbers of deaths and injuries; causes extensive damage or destruction of facilities that provide and sustain human needs; produces an overwhelming demand on State and local response resources and mechanisms; causes a severe long-term effect on general economic activity; and severely affects State, local, and private-sector capabilities to begin and sustain response activities." [3] Hurricanes Katrina and Harvey could fit this description. Hurricane Harvey was the most expensive disaster to impact the US at $125B in damages and losses.

An incident of global significance connotes an incident that transcends national boundaries such as the COVID-19 pandemic which has impacted the entire world. At the end of 2020, it had infected more than 83M and caused more than 1.8M deaths [4, 5].

1.3.3 Post-incident

When an incident is under control, stabilized, and de-escalating, response operations begin transitioning into the recovery phase. The recovery phase is marked by clean-up and restoration. In general, a crisis that can be quickly transitioned from response operations into recovery will have fewer, adverse consequences. But as response operations continue, the incident continues to cause damage within a community. While natural disasters such as Hurricane Harvey could not be controlled, there are some incidents where the duration is determined by incident commander decision-making. Additionally, a community continues to expend and extend its limited public safety resources throughout the response phase. The depletion of and fatigue on resources can adversely impact community resilience and safety in the long term.

During the post-incident phase, operations can move at a slow, steady pace while actions follow general standards for construction, safety, and environmental protection. Unfortunately, those that were affected by the incident may continue to suffer the same effects until their lives, property, and security are restored. While crisis decision-making ends with the transition to recovery, it is important to recognize the importance of and impact from crisis decisions on long-term recovery with respect to infrastructure, economy, social, housing, availability of financial resources, future insurability, and community culture.

Finally, the post-incident phase offers an opportunity to hot wash the incident, identify strengths and opportunities to improve, learn lessons, and ultimately become more resilient. This begins with the hot wash among those engaged in the incident and concludes with an objective after action report and improvement plan. The improvement plan should influence community and organization-wide strategic planning and investments to ensure a more resilient future.

1.4 Identifying Conventional Threats and Hazards

So, where do you start to identify the relevant threats and hazards that could affect your community? A recent FEMA product, the National Risk Index [6], is a great starting point that provides a compilation of location-based natural hazards as well as other information. Information on technological hazards can be more challenging to identify with relevance, but start with local industry, utilities, infrastructure, and information technology-related hazards. Which of these are located in your community? Industrial facilities that store, process, or produce hazardous materials can be found at the EPA Envirofacts website [7]. Power plants (i.e., nuclear, coal, gas, etc.), the electrical network to include substations, water and wastewater systems, natural gas pipelines, transportation networks such as bridges and tunnels, dams, and communication systems such as cell towers, internet lines, and radios represent potential sources of technological hazards. Information on these can be found in local plans or critical infrastructure listings at the emergency management office. Finally, man-made hazards typically include adversarial threats such as active shooters, cyber-attacks, chemical, biological, radiological, nuclear, and explosive hazards. These are harder to predict and estimate the probability of occurrence, but global trends may provide some insight into their relevance.

1.5 Thinking Outside the Box

Traditional threat and hazard identification exercises (such as the Threat and Hazard Identification and Risk Assessment (THIRA)) typically assess about 20–25 types of incidents such as expected natural disasters, relatively common man-made threats, and a few technological hazards. Occasionally, these exercises are informed by catastrophic planning results. But emergency managers should consider non-traditional hazards and threats by engaging experts in various industrial sectors, government agencies such as NOAA, military planners, and federal law enforcement. Future preparedness begins with identifying and considering global incidents and trends. This will guide planners in identifying emerging threats and hazards.

During 2020, we experienced a global pandemic caused by the SARS-CoV-2 virus. In the last two decades there have been several pandemics to include SARS-CoV-1 and MERS-CoV. During the first decade of the twenty-first century, the US extensively funded pandemic planning. The likelihood of another pandemic was substantial.

In 2020, the US has experienced significant civil unrest in most cities. What started as civil rights protests, escalated due to the influence of violent organizations. The tactics used in these riots (e.g., commercial grade LASERs, incendiary devices, and occupy tactics) should provide insight into emerging threats and hazards.

The severity and frequency of natural disasters has increased in some locations. As scientists study climatic cycles, how the climate is changing, and human impact on the environment, leaders should recognize recent trends associated with the severity and frequency of disasters so that we are planning for the future, not just historical incidents. Hurricane Harvey demonstrated this case in point with rainfall that exceeded the 500-year flood.

Technological hazards are also evolving. The Emergency Planning Community Right to Know Act of 1986 requires communities to develop and annually update an Emergency Response Plan (ERP). This ERP identifies the hazardous materials located in and transiting through a community. The mode of transportation changes regularly based on cost, infrastructure, and demand. Fixed facilities adopt new technologies. Economic development organizations attract new industries regularly. These economic activities change the community's risk profile and should be re-evaluated periodically.

Finally, there have been a series of assassination attempts in Russia and the UK. These incidents have unleashed Fourth Generation chemical agents and caused significant impact to responders and the surrounding community. The proliferation of military grade chemical, biological, radiological, and nuclear (CBRN) weapons threaten every city and will require a complex, coordinated response and recovery effort. As rogue nations and terrorist organizations pursue CBRN materials and weapons, we need to recognize that there is some intent to use, not just acquire, these materials. The use of these hazards will only complicate an already complex response. Pairing the right combination of these agents together can exacerbate the consequences and slow the recovery process while consuming extensive resources.

So, what are some ways for leaders to predict novel threats and hazards to their organization? Start with changing conditions and novel situations.

- International news is a great source for identifying new threats among terrorist organizations, geopolitical dynamics, widespread natural disasters, and an outsider's perspective. In recent years, terrorists have modified their methods to include the use of fire, torching, and flammables to destroy land, crops, and deny use complicating peacekeeping situations.
- Geopolitical unrest or conflict can result in changes in economic sanctions and trade restrictions which can affect supply chains. Widespread natural disasters can strain supply chains and demand resources that regional economies take for granted. Recent changes in US oil production and distribution policies resulted in a shortage of domestic supply and increased prices at the gas pump.
- Consider technology advances. For example, the internet of things, apps, machine learning, and smart batteries offer insight into some improvements in technology, but each carries a risk. Hackers have found new ways into networks by hacking unprotected devices on refrigerators. Apps collect information but when hacked, they spill over private information. Machine learning such as the technology used in driverless vehicles can adapt to normal circumstances, but recently studies showed concern over their inability to recognize emergency vehicles.

Electric vehicle batteries when involved in an accident have the potential to experience thermal runaway, meaning traditional firefighting methods with water or foam will not extinguish the fire. The result of thermal runaway is an explosion.

- Economic indicators such as community credit rating, net income, and employment can provide some insight into resident satisfaction with life. Economic pressures can lead to stress, divorce, crime, and contribute to civil unrest.
- Justice and equity in a community or organization represents the balance of power, respect, and trust. When equity is reduced, people become dissatisfied and seek justice. Leaders should gauge their environment for perception of justice and equity. When human dignity is threatened, the potential for crisis looms.

These are several domains to explore where you might identify changes that could affect your context with the potential to emerge into a crisis. Good leadership may be able to avert the crisis from within but look externally to the future threats and hazards: technology, information systems, security situation, politics, legal and regulatory environment, economics (supply/demand), international news, social trends and issues, macro-economic indicators, environmental/climate trends, and global geopolitics.

Summary of Key Points

- Threats and hazards present risks to every community and organization.
- Identifying and assessing the risks in the pre-incident phase can set the conditions for successful outcomes during an incident.
- The spectrum of incidents is comprised of emergencies, crises, disasters, and catastrophes.
- An emergency requires external resources to respond and control the incident.
- A crisis entails a complex, uncertain situation that requires decisive action in order to positively affect the outcome.
- A disaster is a severe occurrence that requires national or international resources to respond and recover.
- A catastrophe results in extremely severe outcomes that adversely impacts a community's economy, mortality rate, and infrastructure; it often requires international resources to begin and sustain a response as local resources are unable to respond.
- Recognizing the type of incident enables decision-makers to marshal resources, establish the appropriate incident command system, and institute the operational tempo for managing the incident.
- Successful crisis management results in a recovery situation, whereas unsuccessful crisis management can result in disastrous consequences.

Keys to Resilience

- Conduct a Hazard Identification and Risk Assessment to determine potential hazards, threats, and their potential impact. You cannot plan for what you do not anticipate.
- Risk-based planning enables leaders to allocate resources based on risk and relative mitigation. Pre-incident planning enables decision makers to prepare for incidents before they occur.
- Prepare crisis leaders with situational awareness, scalable systems, training, and realistic exercises.
- The successful crisis leader makes decisions to accelerate the transition from response to recovery.

Application

In the following scenarios, determine the type of incident based on the definitions in this chapter and describe the type of resources required, including level of government response.

1) A resident reported a fire at a warehouse in the industrial zoned section of a city.
2) An EF-2 tornado destroyed two homes and a playground at 2 pm yesterday.
3) An explosion occurred in the center of a populated city destroying four office buildings at 3 pm.
4) An earthquake registered 7.0 on the Fujitsu scale with the epicenter near a major city.
5) A tsunami destroyed a coastal vacation town.
6) An ammonia pipeline ruptured in a suburban area.

Review the Freedom Industries case study using the Chemical Safety Board's video. https://www.csb.gov/freedom-industries-chemical-release-

1) Discuss or describe the key decisions made or missed during this incident that could have impacted the outcome.
2) What information was needed and when was it needed to make better decisions and transition this incident from response to recovery sooner.
3) Compare the role that each of the following elements could play in building a more resilient community:
   a) Regulations
   b) Communications
   c) Alert and warning systems such as detection technologies
   d) Private infrastructure and operational controls
   e) Zoning
   f) Public notification
4) Construct the system that was needed to successfully manage this crisis and avert a disaster. Consider the following elements of a system: legal/regulatory, incident command system, subject matter experts, training, plans/procedures, strategy, financial.
5) Using the system constructed in response to question 4, apply this system to another historic incident (e.g., using another CSB investigative report, FEMA/State after action report) and evaluate how the elements described in your system could have improved incident response and decision-making.

References

1 Porfiriev, B.N. (1995). Disaster and disaster areas: methodological issues of definition and delineation. *International Journal of Mass Emergencies and Disasters* (November) 13 (3): 285–304.
2 Farazmand, A. (2001). Introduction – crisis and emergency management. Chapter 1 in: *Handbook of Crisis and Emergency Management* (ed. A. Farazmand). New York and Basel: Marcel Dekker, Inc.
3 Federal Emergency Management Agency (1992). Federal response plan with revisions. (FEMA Publication 229).
4 U.S. Centers for Disease Control (2020). COVID-NET: a weekly summary of U.S. COVID-19 hospitalization data. https://gis.cdc.gov/grasp/covidnet/COVID19_5.html (accessed October 2, 2021).

5 The American Journal of Managed Care (2021). A timeline of COVID-19 developments in 2020. https://www.ajmc.com/view/a-timeline-of-covid19-developments-in-2020 (accessed October 2, 2021).

6 Federal Emergency Management Agency (2021). National risk index. https://hazards.fema.gov/nri (accessed October 2, 2021).

7 Environmental Protection Agency (2021). Envirofacts. https://enviro.epa.gov (accessed October 2, 2021).

8 Allinson, R. E. (1993). *Global Disasters: Inquiries Into Management Ethics*. New York: Prentice Hall.

9 Ziaukas, T. (2001). Environmental Public Relations and Crisis Management. In: *Handbook of Crisis and Emergency Management* (ed. A. Farazmund), 245-257. New York and Basel: Marcel Dekker, Inc.

Further Reading

1 Alibek, K. and Handelman, S. (1999). *Biohazard: The Chilling True Story of the Largest Covert Biological Weapons Program in the World – Told from Inside by the Man Who Ran It*. New York: Random House.

2 Director of National Intelligence (2021). Annual threat assessment of the intelligence community. https://www.dni.gov/index.php/newsroom/reports-publications/reports-publications-2021/item/2204-2021-annual-threat-assessment-of-the-u-s-intelligence-community (accessed October 2, 2021).

3 Federal Emergency Management Agency (2021). National preparedness annual report. https://www.fema.gov/emergency-managers/national-preparedness (accessed October 2, 2021).

4 The Intel Center (2021). www.intelcenter.com (accessed October 2,2021).

5 Suskind, R. (2007). *The One Percent Doctrine: Deep Inside America's Pursuit of Its Enemies Since 9/11*. New York: Simon and Schuster.

Key Terms Used in This Chapter

Threat: Natural, technological, or human-caused occurrence, individual, entity, or action that has or indicates the potential to harm life, information, operations, the environment, and/or property.

Hazard: Something that is potentially dangerous or harmful, often the root cause of an unwanted outcome.

Incident: An occurrence, natural or human-caused, that requires a response to protect life or property. Incidents can, for example, include major disasters, emergencies, terrorist attacks, terrorist threats, civil unrest, wildland and urban fires, floods, hazardous materials spills, nuclear accidents, aircraft accidents, earthquakes, hurricanes, tornadoes, tropical storms, tsunamis, war-related disasters, public health and medical emergencies, and other occurrences requiring an emergency response.

Emergency: Any incident, whether natural, technological, or human-caused, that requires responsive action to protect life or property. Under the Robert T. Stafford Disaster Relief and Emergency Assistance Act, an emergency means any occasion or instance for which, in the determination of the President, Federal assistance is needed to supplement State and local efforts and capabilities to save lives and to protect property and public health and safety, or to lessen or avert the threat of a catastrophe in any part of the United States.

Disaster: An occurrence of a natural catastrophe, technological accident, or human-caused event that has resulted in severe property damage, deaths, and/or multiple injuries.

https://training.fema.gov/programs/emischool/el361toolkit/glossary.htm#A

Catastrophic Disaster: An event that results in large numbers of deaths and injuries; causes extensive damage or destruction of facilities that provide and sustain human needs; produces an overwhelming demand on State and local response resources and mechanisms; causes a severe long-term effect on general economic activity; and severely affects State, local, and private-sector capabilities to begin and sustain response activities. Note: The Stafford Act provides no definition for this term. ([3], FRP Appendix B).

Crisis: "a decisive or critical moment or turning point when things can take a dramatic turn, normally for the worse" [8].

Crisis: "Crises involve events and processes that carry severe threat, uncertainty, an unknown outcome, and urgency … Most crises have trigger points so critical as to leave historical marks on nations, groups, and individual lives. Crises are historical points of reference, distinguishing between the past and the present. … Crises come in a variety of forms, such as terrorism (New York World Trade Center and Oklahoma bombings), natural disasters (Hurricanes Hugo and Andrew in Florida, the Holland and Bangladesh flood disasters), nuclear plant accidents (Three-Mile Island and Chernobyl), riots (Los Angeles riot and the Paris riot of 1968, or periodic prison riots), business crises, and organizational crises facing life-or-death situations in a time of rapid environmental change. … Crises consist of a 'short chain of events that destroy or drastically weaken' a condition of equilibrium and the effectiveness of a system or regime within a period of days, weeks, or hours rather than years. … Surprises characterize the dynamics of crisis situations … Some crises are processes of events leading to a level of criticality or degree of intensity generally out of control. Crises often have past origins, and diagnosing their original sources can help to understand and manage a particular crisis or lead it to alternative state of condition" ([2], pp. 3–4).

Crisis: "a hard and complicated situation … or a turning point – a decisive crucial time/event, or a time of great danger or trouble with the possibilities of both good and bad outcomes" ([1], pp. 291–292).

Crisis: "a situation that, left unaddressed, will jeopardize the organization's ability to do business." [9].

Crisis Management: "Key to crisis management is an accurate and timely diagnosis of the criticality of the problems and the dynamics of events that ensue. This requires knowledge, skills, courageous leadership full of risk-taking ability and vigilance. Successful crisis management also requires motivation, a sense of urgency, commitment, and creative thinking with a long-term strategic vision. In managing crises, established organizational norms, culture, rules and procedures become major obstacles: administrators and bureaucrats tend to protect themselves by playing a bureaucratic game and hiding behind organizational and legal shelters. A sense of urgency gives way to inertia and organizational sheltering and self-protection by managers and staff alike. … Successful crisis management requires: (1) sensing the urgency of the matter; (2) thinking creatively and strategically to solving the crisis; (3) taking bold actions and acting courageously and sincerely; (4) breaking away from the self-protective organizational culture by taking risks and actions that may produce optimum solutions in which there would be no significant losers; and (5) maintaining a continuous presence in the rapidly changing situation with unfolding dramatic events." ([2], p. 4)

Hot Wash: A facilitated discussion held immediately following an exercise among exercise players from each functional area that is designed to capture feedback about any issues, concerns, or proposed improvements players may have about the exercise. The hot wash is an opportunity for players to voice their opinions on the exercise and their own performance. This facilitated meeting allows players to participate in a self-assessment of the exercise play and provides a general assessment of how the jurisdiction performed in the exercise. At this time, evaluators can also seek clarification on certain actions and what prompted players to take them. Evaluators should take notes during the hot wash and include these observations in their analysis. The hot wash should last no more than 30 minutes.

Additional terms and definitions can be found at the following source:

Federal Emergency Management Agency (2006). Hazards, Disasters and US Emergency Management: An Introduction. https://training.fema.gov/hiedu/docs/hazdem/appendix%20-%20 select%20em-related%20terms%20and%20definitions.doc#:~:text=Crisis%3A%20 %E2%80%9C%E2%80%A6,and%20property%20of%20its%20members%E2%80%A6 (accessed 2 October 2021).

2

Operating Contexts – Command and Coordination

In Chapter 1, we developed a context for the spectrum of incidents and explored emerging threats and hazards which could impact future response actions and decision-making. In this chapter, we are going to explore the organizational context along the spectrum of incidents. Single organizational operating contexts present their own challenges, but when combined with different jurisdictions and public–private interests, the organizational challenges become much more complex and political. Therefore, it is not only important to understand how the scale of an incident and the multitude of organizations involved in the response are supposed to interact, but also the practical issues encountered. This foundation is essential to preparing an organization or community for an incident so that these issues can be explored with response partners and aid agreements can be constructed in a way to minimize friction and optimize cooperation and coordination.

Learning Objectives

At the end of this chapter, you will be able to:

- Identify sources of dissonance between various groups that respond to threats and hazards.
- Understand the significant challenges to interoperability.
- Understand the importance of engaging both the private and public sector in emergency response, annual exercises, and pre-incident planning.
- Construct a unified command system.

2.1 Public Safety Context

Let's begin with the public safety context. In 2004, the Department of Homeland Security and FEMA published the National Response Plan, which has been superseded by the National Frameworks, and the National Incident Management System (NIMS). NIMS has several components, one of which is the Incident Command System (ICS). ICS is the standard organizational structure that is implemented during an emergency and is shown in Figure 2.1. The ICS is flexible and scalable (i.e., only those positions which are necessary should be filled and it can increase or decrease in size as needed). The incident command system is established at the onset of an incident by the first responder on scene, who becomes the de facto incident commander. As responders arrive, a qualified incident commander (IC) will assume command and assign roles to responders as needed to ensure unity of command; this is a NIMS guiding principle stating that "each

Crisis-ready Leadership: Building Resilient Organizations and Communities, First Edition. Bob Campbell, PE.
© 2023 John Wiley & Sons, Inc. Published 2023 by John Wiley & Sons, Inc.

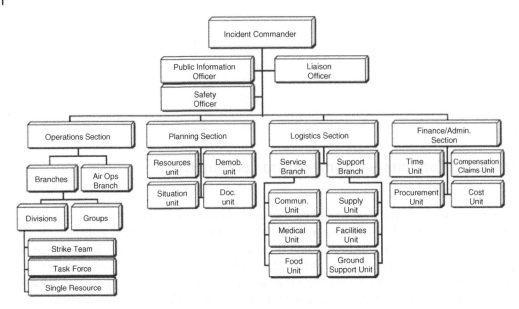

Figure 2.1 Incident command system. *Source:* Federal Emergency Management Agency (May 2008). Incident Command System Training Review Material.

individual involved in incident management reports to and takes direction from only one person." [1] The IC also establishes incident objectives for execution by the Operations Section. As the incident grows, a Planning Section assists the IC with development of incident objectives for the next operational cycle. The Logistics Section handles the growing number of transportation, housing, food, care, personnel support, and shelter needed to support the response. The Administration/ Finance Section tracks expenses, allocates charges, and provides general administrative support to responders. These Sections comprise the General Staff.

The Command Staff provides support to the IC and interacts externally with the public and other entities. For example, the Liaison Officer coordinates operations with external agencies such as State, Private and Federal agencies. The Public Information Officer provides information to the media and public. The Safety Officer advises the IC on scene health and safety issues to ensure responders are safe from hazards.

This standardized approach to incident response transcends public safety organizations and jurisdictions, but sometimes will vary in how it is implemented due to limited resources in a community (e.g., one person may fill several roles). In order to qualify for federal grants, communities are required to adopt the ICS component of NIMS. There is a growing body of knowledge and qualification system which identifies responsibilities for each position, outlines training and qualifications for those assigned to each role, and job action sheets to aid those who are filling each role.

Establishing an Incident Command during an incident is very important to ensure a response is successfully completed. Without an organized command and control system like the Incident Command System, operations tend to overwhelm responders. While national policy requires the use of the ICS under NIMS, not all entities in a community employ the ICS; however, many organizations have some type of system in place for response. Additionally, it is important that all entities whether public or private work together to accomplish the incident objectives through proper allocation of resources. This may require a Unified Command to accomplish the objectives when there are multiple agencies or entities involved in the response to ensure it is effective.

According to NIMS, "incident command" is responsible for:

1) Tactical activities to apply resources on the scene;
2) Incident support through operational and strategic coordination, resource acquisition, and information gathering, analysis, and sharing;
3) Policy guidance and senior-level decision-making;
4) Outreach and communication with the media and the public to keep them informed about the incident.

> According to NIMS, the **incident command system** (ICS) is, "a standardized approach to the command, control, and coordination of on-scene incident management that provides a common hierarchy within which personnel from multiple organizations can be effective."

Today, many agencies are finding that a mobile transport incident command or mobile command post is a must-have for a level of situational awareness that typically only comes with being on-scene. With improvements in mobile technology, these command posts allow for capabilities not feasible a decade ago and the ability to quickly shift the location based on the changing characteristics of the response. These types of emergency events often include a multi-agency response that can include personnel from local, state, and federal response agencies. The mobile transport incident command can provide the incident commander with the tools needed for communication, coordination, and control. This improves situational awareness on scene with mobile technology, and provides the ability to move to another area, while coordinating resources.

2.2 Private Sector Context

The private sector plays many roles in emergency preparedness. While 85% of the US critical infrastructure is owned by the private sector, all private organizations play some role in a community (e.g., employer, generation of tax revenues, value-added products and services, transporter of goods, banking, education, mining and agriculture provides primary inputs to the economy, construction and manufacturing builds infrastructure and products, retail delivers goods to end users, etc.). Therefore, all organizations have some stakeholders who rely on their products and services. This drives a preparedness requirement. Additionally, some organizations are more likely to interface directly with the public sector during a crisis because of the intrinsic value or hazard that they present within a community. For example, critical infrastructure and congregant facilities offer certain value to specific populations within the community. These may include hospitals, nursing homes, prisons, and schools. Whereas chemical storage, processing, and transportation facilities present upstream value to manufacturers and end users, but present a localized risk from potential hazardous material (HazMat) incidents. Throughout the remainder of this chapter, we will more thoroughly explore the operating context between operators of HazMat facilities and the public sector since this represents a more complex, public–private sector interface where as other cases of public–private sector interfaces are coordinated through a single liaison or on-going relationship. Regardless of the involvement and scale of the public–private sector interface, this chapter will address a variety of operational coordination issues that may be encountered.

The private sector is encouraged but is not required to adopt the same public sector ICS. While there are several environmental, health, and safety regulations which require an incident command system during hazardous material spills, the details on the ICS are incomplete. Many private sector organizations adopt a logical system that fits their organizational structure and operations. In many cases the organizational structure matches the public sector ICS at a high level, but becomes more organization-centric at lower levels of the organization. The exceptions are large, multinational companies which centrally manage an incident through the deployment of resources to the incident location (e.g., oil and gas companies, rail companies, etc.). The next two subsections will explore stationary and mobile hazards that are operated by the private sector and the implications for coordination with the public sector.

2.2.1 Stationary Hazards

Small companies may simply offer a liaison to the public sector responder and defer to public responders to support their incident. Medium size companies may have some response capabilities but rely on a partnership with the public sector to resolve large incidents beyond their limited capabilities. Large companies may have local resources but usually retain reach back, deployable resources to support response and can in most cases handle response operations on their own (e.g., Power companies). In this case, the public sector oversees the response operations to ensure compliance and safety while providing essential information to the public. These roles can vary depending on the size and scope of an incident as outlined in Table 2.1.

There is no requirement for private companies to adopt the FEMA ICS or even establish a response organization; it is merely "recommended" under the Hazardous Waste Operations and Emergency Response, or HAZWOPER regulations. OSHA requires an emergency action plan to protect employees from various types of incidents. While companies are not required to have a response organization, they are responsible for any incidents, environmental releases, and the health and safety of their workers. This is why it is important for the community preparedness planners and first responders to coordinate with private sector facilities to understand their Emergency Action Plan (EAP) and any emergency response plans that may be required under various environmental regulations, such as the Clean Air Act/Risk Management Program (CAA/RMP), Clean Water Act/Spill Prevention Control and Countermeasures (CWA/SPCC), Clean Water Act/Storm Water Pollution Prevention Plan (CWA/SWPPP), Resource Conservation and Recovery Act (RCRA), hazardous waste management plans, and hazardous materials management plans. The type of response plan that may be in place could vary based on regulatory requirements,

Table 2.1 Variations in ICS elements and resources by size of the organization.

	Small Organization	Medium Organization	Large Organization
ICS elements	Site manager	Site manager Response teams FEMA-consistent ICS Facility-specific ICS	Emergency Communications Center Crisis management Coordination center Incident management teams Company-specific ICS
Resources	Call 911	First aid, EMT Site security Fire brigade Hazmat team Public information officer Contracted cleanup	Deployable response teams

size of company and scale of potential incidents. Nonetheless, private industry that stores or processes hazardous materials has the potential to cause an impact on the surrounding community and require public sector responders to support and protect the community.

There are several issues to collaborate on so that both the private and public responders are able to demonstrate operational control.

2.2.2 Mobile Hazards

For mobile hazards such as those transported over rails, pipelines, barges and roadways, the response from the conveyor will likely be delayed until the response team can deploy and arrive on scene.

So here are two potential scenarios:

1) The private company initiates an emergency response plan and is joined later by the public sector first responders;
2) The public sector initiates an emergency response plan and is joined later by private sector-sponsored response/recovery teams.

Notification procedures are implemented following the report of a HazMat incident. All incidents in the rail community are considered an emergency until identified as otherwise. Rail companies have developed an emergency response plan that includes their steps in a response process. During an incident the notification procedures start with the dispatcher, trainmaster, or yardmaster. They will obtain as much information as possible about the incident and notify the service interruption desk in the operations center. Rail companies utilize their Response Management Communication Center (RMCC) where dispatchers receive calls from the public, law enforcement, and others reporting emergencies and incidents along their rail network. RMCC follows all regulations regarding notification of local, state, and federal agencies in the event of an accident and works closely with first responders throughout an incident. The communications center will then notify the Hazardous Material Environmental Response (HMRs), US/Canadian/State/Provincial/Industry agencies such as the National Response Center, Nuclear Regulatory Commission, Center for Disease Control, Association of American Railroad Bureau of Explosives, Transport Canada, and State/Provincial agencies, the shippers and customers, and the resource operations center. The HMRs are responsible for contacting and providing follow-up information to HazMat response personnel, state/federal regulatory agencies, and shippers.

Depending on the levels of incidents, the rail company has an emergency response plan that involves different levels of responsibilities and actions. The ICS is implemented to effectively control an emergency situation.

The first person to arrive on-scene will classify the incident based on the severity; this is also known as scene size up. There are three levels of incidents:

- Level 1 Incidents:
  - No evacuation.
  - First responders can control release without support.
  - HMRs can manage and mitigate release.
- Level 2 Incidents:
  - Release or potential release is effectively contained with equipment/supplies.
  - Evacuation need only in immediate area.
  - Government agencies respond with specialists.
- Level 3 Incidents:
  - Releases not properly abated with HMRs and equipment.
  - Evacuation required.
  - Multiple governmental agencies involvement.

2.3 Combined Public–Private Operating Context

Given the variations in the scale of an incident and the private sector response operations involved, it is critical for the public and private sector organizations to conduct pre-incident planning so that (1) the interface between the organizations can be defined and (2) the need for resources can be determined.

Using the Spectrum of Incidents as a guideline, the organizations should determine the capabilities of the private organization to handle the response. There are two components to this pre-incident planning analysis: (1) the type of incident and (2) the scale of an incident. Using Table 2.2 below, let's consider a few examples.

To better illustrate the context for response and spectrum of incidents, let's review a few case studies using some similar hazards but various levels of response resources among levels of government and the private sector.

2.3.1 El Dorado Chemical Company Emergency Case Study

In 2009, the El Dorado Chemical facility in Bryan, TX caught fire as a result of a welding incident. This facility stored tons of ammonium nitrate fertilizer which had the potential to catch fire and detonate. So, when the fire department arrived on scene and recognized the risk, they evacuated 72,000 people within an eight-mile radius of the incident and allowed the fire to smolder and burn itself out instead of applying water that could have caused it to spread. This decision was made in part by firefighters being alerted to the dangers presented by El Dorado's hazard reporting and previous history with fires at other fertilizer storage facilities. Nonetheless, public and private sector representatives had coordinated and planned together prior to the incident to ensure coordinated response. However, according to after-action report, the fire chief of Bryan reported that this incident provided some lessons learned in communication between the incident command on-scene and the emergency operations center. This case demonstrates the need for pre-incident planning and coordination to ensure safe outcomes.

Table 2.2 Spectrum of incidents.

Incident Type	Incident	Emergency	Crisis	Disaster
Explosion	Small propane tank	Fuel storage tank with fire	Facility-wide fire with large scale chemical process	Ammonium nitrate storage facility explosion
Chemical spill	55-gallon chemical spill	Ammonia pipeline rupture	Large chemical release with potential to react with other substances	Ongoing chemical leak affecting drinking water source and evacuations
Resources	POL spill kit	4 Hazmat technicians, patch kit, detectors	Remote/auto-shutoff valves. Incident management team. AFFF foam	Contract hazmat response and recovery teams
Scale	Up to 500 gallons	Single entry/hour	20-person incident management team: 10 hazmat techs, decon team, detection, 3 EMTs, fire brigade (5). 500 lbs foam	Financial resources and insurance needed to support response and recovery operations

2.3.2 West Fertilizer Company Crisis Case Study

On April 17, 2013, the West Fertilizer Company which stored 30–50 tons of fertilizer grade ammonium nitrate caught fire. Within 20 minutes of the first call to the emergency communications center, the facility exploded killing 12 first responders and three members of the public. The blast destroyed 150 buildings, injured 262 people, and caused $230 M in direct economic damages to the community. Sadly, investigators from the Chemical Safety Board found that initial responders did not operate under an incident command system, few were trained to respond defensively to an incident of this magnitude, none were trained to respond offensively, a pre-incident plan did not exist, and prior hazardous material planning was limited to the anhydrous ammonia tanks on site, but not the ammonium nitrate. The subsequent response and recovery entailed multi-jurisdiction, multi-agency response to conduct search and rescue, transport casualties to hospitals, and coordinate with federal agencies that sought to investigate the cause of the fire. This case illustrates a crisis that could have been averted with proper training specific to the hazards and an understanding of proper response procedures for an incident of this magnitude. For more information on the subsequent response and recovery, read the profile in crisis leadership on Frank Patterson in Chapter 10. Frank served as the incident commander and led this response to its completion and into recovery operations.

2.3.3 Superstorm Sandy Disaster Case Study

On October 29, 2012, Hurricane Sandy battered the east coast of New Jersey and New York with heavy rains, strong winds, and record storm surges taking 162 lives, causing tens of thousands of injuries, and displacing hundreds of thousands of residents. New York City's after action report summarized the themes of lessons learned as [2]:

- Improved evacuation included updated evacuation zones and improved public notification messages on how to protect themselves;
- Improved accessibility to storm-related information and services especially those with functional and access needs;
- Better integration of data across the City's information platforms and agencies to increase situational awareness;
- Additional capacity to respond to large-scale incidents to include pre-incident planning and identification of resources needed for restoration and better coordination with private building owners;
- Better coordination of relief to include deployment of volunteers and donations to residents and building owners;
- Development of mid- to long-term housing plans for displaced residents;
- Partnership with federal, state, utilities, and private companies that provide essential services.

The FEMA after-action report echoed the "notable challenges in how FEMA coordinates with its Federal partners, supports state and local officials and disaster survivors, integrates the whole community, and prepares and deploys its workforce." [3] While FEMA has made significant improvements in these areas over the last decade, this case highlights the challenges of operational coordination among different agencies in large, complex disasters.

2.3.4 COVID-19 (Global Incident of Significance) Case Study

The COVID-19 pandemic struck the US in February 2020 and quickly spread with multiple waves which still continue as I write this book. The global significance of this disaster has had far-reaching effects which have stressed multiple aspects of our emergency management and public health communities. Traditionally, the public health community has been designated as the lead federal agency for pandemic planning. The same resonates at the state and local levels. But as we witnessed with the Ebola Virus Disease incident in Texas during 2014, the emergency management coordinators joined the health department directors in a unified command role in both Texas and Virginia (home of Dulles Airport which received international flights from the affected region). However, not long after the COVID-19 pandemic emerged in the US, the public health departments at the local, state, and federal level were overwhelmed and required emergency management agencies to step in to play a more prominent role. For more details on a local example, read the profile in crisis management with Chad Hawkins in Chapter 10. According to the FEMA initial assessment report, the President and Vice President informed the FEMA Administrator that FEMA would be leading the response five days after the President had announced that HHS would serve as the Lead Federal Agency consistent with the Pandemic Crisis Action Plan which was finalized in March 2020 [4]. As FEMA took the lead, they faced numerous challenges integrating task forces into the existing National Response Coordination Center structure. This resulted from lack of NIMS/ICS adoption from other federal agencies, state, local, tribal, and territorial partners. FEMA also established the National Joint Information Center, but the message approval process was complex due to both the White House's role as well as the need to coordinate with 40 departments and agencies involved in operations.

These cases illustrate how the operating context can grow in complexity among external partners and stakeholders as the scale of the incident grows. This underscores the importance of a whole of community approach to planning and preparedness. Even the most robust plans can reach breaking points as we have witnessed with COVID-19.

During public health emergencies, the incident command structure varies from traditional incident command as the coordination role grows in magnitude. In some cases, the Emergency Operations Center (EOC) becomes the focal point for coordinating resources and establishing operations. In other cases, the incident command system is utilized at a high level while individual sites utilize more detailed ICS structures. During COVID, the EOC was most widely utilized to provide resource support where needed. Public Health Departments utilized the ICS, but decision-making occurred in coordination with the Policy Group (e.g., elected officials) and other partners. In most cases, the state or Governor's office was the key decision maker. This elevated the role of the state health department superseding local decision-making. Here are a few examples of ICS. Under the traditional model of emergency management, incidents are locally executed, state managed, and federally supported. But during the COVID-19 pandemic, the state level of government took the lead in establishing executive orders and public health directives rather than local health departments, although there were some exceptions. The federal government supported states with financial resources and guidance which became de facto rules adopted by the states, specifically those states with emergency temporary standards that were predicated on the more stringent of state mandates and federal guidance.

The local EOC still played an important role in monitoring health-care facility capacities, incident rates, and partnering with business councils to support economic development initiatives such as relaxing permitting rules to allow restaurants to utilize outdoor seating on sidewalks and streets. Early in the pandemic, public health and health care dominated the response and messaging. But as the secondary effects on the economy emerged, the state EOC played more prominent roles in solving state-wide issues such as economic develop organizations sourcing parts, networking the business community to solve supply chain issues, and pivoting the manufacturing base to make ventilators and PPE.

2.4   Escalating Scale of Incidents Lead to Multijurisdictional Response

As the scale of an incident increases beyond local response capabilities, the local jurisdiction will likely activate any mutual aid agreements with surrounding jurisdictions. This could be a limited request for fire department support or a regional HazMat team, but it could also entail more extensive resources to support ongoing EOC staffing or incident management teams. Once mutual aid is exhausted, the local jurisdiction may request assistance through the state who may draw upon state agency resources such as the national guard, state police, or state-run regional incident management teams. When the scale of the incident exceeds state resources, the state may request resources from the federal government.

The benefit of this system is that all organizations have adopted NIMS and will operate under the common ICS. FEMA has defined resources using the "resource-typing" system which lists resources in a catalogue format specifying the type and quantity of resources with some description of what can be expected. This has facilitated the request and provision process. Once the external resources arrive, the IC can easily integrate the team into the ICS so that unity of command continues.

2.5   Challenges of Multijurisdictional Response

There are several challenges with multijurisdictional response operations. First, the external resources are only available to the extent agreed upon in the Mutual Aid Agreement. These are not necessarily free resources. Therefore, the IC may not have full use of the resources. Second, distance affects arrival time and delays the response. This is particularly true in rural areas. Third, organizational structures don't easily allow for the utilization of individuals but rather teams. So, as more teams arrive, there may be more bureaucracy to navigate. In the case of the West Fertilizer Company, the incident commander was from a neighboring jurisdiction. As the IC, he was committing resources for the City of West and making decisions on behalf of the City, but he was not employed by the City. This required an extra layer of coordination with the Mayor.

When coordinating with the State or Federal government, local incident commanders can become quickly overwhelmed with the myriad of resources if the ICS structure does not keep up with the expansion of the response effort. While state and federal resources play a supporting role, organizationally, they are still accountable and under the control of the state or federal government leaders. This has led to miscommunication and confusion during previous incidents such as the Freedom Industries incident where the company released chemical substances into the Elk River, contaminating the drinking water source. As experts at all levels debated the safety of the drinking water, there were various public positions from the company, mayor, and CDC. Similar confusion occurred during the Ebola crisis when hospital staff and administrators disputed proper protocols and safety measures, as the President and the CDC weighed in on their opinions which conflicted with local and state positions. Public information consistency becomes much more difficult to unify across the local, state, and federal government. Figure 2.2 outlines several of the dominant forces and elements encountered in multi-jurisdictional responses.

2.6   Large Scale Disaster, Multinational Context, and Challenges

Similar challenges related to authority, support scope, and politics arise during multinational response efforts. Each entity with their own agenda may compete for resources or messaging making multinational response a very complex operating context.

Mutual Aid	State	Federal	International
• Lack of control • Limited financial resources • Delayed response	• Incident management team construct • Integration with local ICS	• Resources report to federal government • Coordinating lines of effort • Control and communication	• Political • Economic

Figure 2.2 Dominant forces in multijurisdictional response.

It was not long after the Beirut Ammonium Nitrate explosion that killed 200 people and damaged thousands of structures, that the power struggle among the people and elected officials became evident. Calls for the Prime Minister to resign were effective. International influence from outside governments positioned themselves to weaken the remaining government. As US experts arrived to assist with the investigation, it became evident that some in the Lebanon government did not want to properly investigate as the blame might point back at the bureaucracy that seized and impounded the cargo but failed to store or dispose of it properly. Political forces create the greatest challenge in this context.

2.7 Multidisciplined Response

Multidisciplined response operations also present several challenges to incident command. First, culture clashes over responsibility and process can impede response operations. Second, lack of trust based on history, personality clashes, and independent thinking have created challenges to integrating disciplines. Third, stove-piped organizational mentality prevents organizations from achieving unity of effort. Fourth, different operating standards and credentialing prevent integration of disciplines from trusting each other's competence and proficiency to perform some tasks.

A clear understanding of roles and responsibilities as outlined in plans can help address these challenges. By defining roles and responsibilities during the pre-incident phase, incident commanders can become more familiar with the resource capabilities and limitations associated with each response agency. Practicing multidisciplined response operations by training together and exercising responses can help smooth over many of these challenges from different organizations, improve trust, and integrate at appropriate interfaces.

The following technical examples illustrate some of the of interoperability challenges for multi-discipline, multi-agency response operations.

2.7.1 Credentialing and Certifications

This becomes an issue when determining the level of care provided by EMTs on site and public sector EMS. In some cases, treatment can be delayed due to uncertainty of the reliability of the assessment performed at a private facility and whether personnel are credentialed or certified to an acceptable level by the public sector.

2.7.2 HazMat Response Training and Certification

Each state typically establishes training and certification standards in accordance with NFPA for HazMat responders. Many HazMat responders are required to get certified through an IFSAC accredited state training facility. This is not the case for private sector HazMat teams who may pursue ProBoard certification or perhaps simply meet the EPA and OSHA requirements outlined under HAZWOPER (29 CFR 1910.120) for emergency response.

2.7.3 Consensus Standards for Operations

Consensus standards for operations could affect a responder's performance, practices or operations. Some examples are NIMS, ICS, National Fire Protection Association (NFPA), ISO.

2.7.4 Methods of Triage, Decontamination, and Sampling

It may be helpful to discuss these in planning, train together for consistency and practice during exercises to ensure interoperability. There are several different standard processes for triaging casualties. The local standard should be pre-determined and communicated to any response agency and receiving hospital. Casualty decontamination is important to protecting the infrastructure of the ambulance and receiving hospital, as well as those providing care during transport and at the receiving facility. Coordinating acceptable standards for decontamination by assigning roles, responsibilities, and employing standard methods, can ensure swift, effective decontamination, transport, and entry into a receiving facility where timely treatment can be provided. This is only possible if everyone agrees on what is acceptable.

2.7.5 HazMat Detection

Responders have different technology, especially when it comes to HazMat detection. HazMat detectors are not created equally. Some variances include limits of detection, principles of operation, the limits of quantification, effective calibration, and intrinsic safety. Incident commanders may be making decisions based on sampling and analysis of results from hand-held detectors; therefore, they need to know that they can rely upon the results. When unfamiliar technology/ detectors are introduced, this could cause confusion with respect to units of measure, interpretation of results, action levels, and acceptable risks.

2.7.6 Public Communication

Each organization will likely have a public information officer with different motives and expectations. For example, private sector organizations may emphasize brand and reputation protection over sharing negative information. Likewise, public sector public information officers (PIOs) may emphasize messages based on political leader influence and public protection. If the PIOs do not collaborate on message development and delivery, there is a potential for conflicting or confusing messages that result in ineffective public protective action implementation. While their respective interests and stakeholders may be different, they should work together in a joint information center to develop consistent messages related to protection of the public.

2.7.7 Technical Specialists

Since technical incidents will require technical specialization, each party may seek an advisor or outside consultant to inform them about potential risks. Sometimes the advisor is from another government agency. It is important to collaborate between the public and private entities along with these advisors to ensure factually accurate information is shared. While a private sector company should know their operations and processes better than most outside consultants, a collaborative approach to gathering information and determining risks should be utilized to achieve faster results.

2.7.8 Recovery Contractors

In some cases, companies may rely on third-party response or recovery contractors. It is important for public responders to understand this part of the private company's plan so that they can ensure timely actions are taken to address public safety risks. Likewise, some public entities rely on private HazMat response and recovery entities to supplement public responders. Additional collaboration and information sharing is needed to ensure effectively communicated assumptions on which entity is responsible, which entity is expected to respond, and how these entities will coordinate to ensure a timely response and transition to recovery.

2.8   Unified Command

There may be cases where incidents require the IC to share command with another agency or private company. In these cases, the IC will need to consider several factors on how they will work together with another entity. It is important to coordinate incident response roles and responsibilities during the planning phase. There are also several challenges to integrating respective command systems into the unified command.

Credentialing and certifications become very important in a unified command. All personnel need to have the same credentials and certifications, or they need reciprocity to mitigate legal risks and ensure consistency in standards and operations.

ICS positions, roles, and responsibilities are scalable. So, when implementing a Unified Command, those involved should ensure clarity of roles and responsibilities, not assuming that personnel from the other entity will interpret those the same as personnel from their organization. It is important to pre-plan to determine who will be in charge and who will play what roles during an emergency. Job action sheets can aid in ensuring interoperability. Response equipment among different agencies is not always compatible and interoperable. Also, organizations may have different standards for maintenance, calibration, and operability. When sharing or co-utilizing equipment, recognize that different models of the same equipment may have different capabilities, limitations, and features such as software. It is crucial to understand these differences before employing the equipment, especially with more sophisticated equipment such as chemical detectors, as misinterpretation of results can lead to the wrong decisions.

Response can become a challenge if plans are not developed together or coordinated. Pre-incident planning is very important when working across agencies. A coordinated pre-incident plan will help each agency define and agree on roles prior to an incident.

Communications can also be a challenge during an incident. If the response includes a private and public sector entity, the lines of communications may not be consistent. Different operating frequencies and communications equipment can also hamper effective communication. Furthermore, common terminology that is adopted under NIMS may not be used in the private sector.

Finally, multi-agency responses can evolve into a parking lot of mobile command vehicles that don't effectively integrate under the ICS. These mobile resources need to be used to support operations, not a barrier to command, control and communication.

Summary of Key Points

- The Incident Command System (ICS) is a standardized approach to the command control, using this system, responders can be more efficient in their response. Additionally, it establishes a hierarchy which makes communication during an incident easier to navigate.
- During a disaster, a Public–Private Response is essential to use all of the available resources. It is important to include both the public and private sectors in the pre-incident planning phase.
- It is important to identify incident response roles and designate responsibilities across multiple agencies in order to establish Unified Command.
- Engaging state EOC and regional IMTs creates a Multi-jurisdiction Response. This kind of response should always be a priority at large-scale exercises or planning.
- Engaging organization that have different specialties and backgrounds produces a Multidiscipline Response. This kind of response is important for delegating responsibilities to experts and establishing unified command.
- There are several Key Challenges to Interoperability (such as: varying levels of certifications and response methods). Despite these challenges, interoperability is crucial to improving response to threats and hazards.

Keys to Resilience

- Pre-incident planning is essential to sharing information and pre-determining initial actions based on incident analysis.
- Ensure plans clearly identify roles, responsibilities, and processes for responders.
- Multi-agency responders should train and exercise together regularly.
- Annual large-scale exercises should engage multi-jurisdiction response by activating mutual aid agreements, engaging the state EOC and regional IMTs.
- Local Emergency Planning Committees should facilitate planning discussions, tabletop and functional exercises with high-risk chemical facilities and transporters in their community. A three-to-five-year cycle should be developed to engage at some level.
- Public sector agencies should coordinate with vulnerable facilities and populations to ensure they have plans and are aware of public sector capabilities and limitations in providing support to various types of incidents.

Application

1) Review the video case study from the CSB ("Emergency Preparedness" from CSB 8:20-9:22 min). Discuss potential disconnects between facilities and responders, along with how IMS and ICS play a role in bridging those gaps. https://www.csb.gov/videos/emergency-preparedness-findings-from-csb-accident-investigations
2) Read the after-action reports referenced throughout this chapter. Identify some of the common incident command and coordination issues that are common. Search for local after-action reports from incidents that impacted your community and compare the command and coordination findings with the major incidents cited in this chapter.
3) Attend your LEPC meeting (public is invited according to EPCRA) and review your Emergency Response Plan (ERP) for your district. Assess the level of public/private preparedness based on the following:

- Representation of both public and private sector organizations at the meeting and in the ERP.
- Potential barriers to integrating during a response.
- Limited resources in the community which would require external resources.

4) Interview a local hospital emergency manager to determine whether their Hazard Vulnerability Assessment addresses natural disasters, technological hazards, and man-made incidents discussed in Chapter 1.

5) Examine your incident command system and identify how those within your organization will relate to external entities such as other local jurisdictions, state, federal, and private partners such as utilities.

6) This is a group exercise that is best conducted in the context of a class with multiple organization represented from both the public and private sector to provide input. Participants may be divided into groups to discuss their respective ICS roles. The goal is to construct an incident command chart for your organization and include lines of coordination with external organizations. This may also include unified command to the extent applicable.

Scenario

A major natural disaster strikes your organization and community. The effects of the disaster include power outages, limited communications such as cell signals and no internet, loss of water pressure, facility damage, production disruption, overwhelmed hospitals, and increased safety risks to personnel in the facility and throughout the community.

- Construct your organization's ICS chart. Identify who is assigned to various roles.
- Identify external stakeholders and how you plan to coordinate with them.
- What capabilities and capacities does your organization have to respond to this disaster?
- What capabilities and capacities do your partner/external organizations have to support?
- What are some of the challenges that you anticipate? Discuss them with your team.
- What are some potential solutions to these challenges?

References

1 Federal Emergency Management Agency (2017). National incident management system. https://www.fema.gov/sites/default/files/2020-07/fema_nims_doctrine-2017.pdf (accessed October 2, 2021).

2 New York City. (2013). NYC Hurricane Sandy after action report. https://www1.nyc.gov/assets/housingrecovery/downloads/pdf/2017/sandy_aar_5-2-13.pdf (accessed October 2, 2021).

3 Federal Emergency Management Agency (2013). Hurricane Sandy FEMA after action report. https://s3-us-gov-west-1.amazonaws.com/dam-production/uploads/20130726-1923-25045-7442/sandy_fema_aar.pdf (accessed October 2, 2021).

4 Federal Emergency Management Agency (2021). Initial assessment report key findings and recommendations https://www.fema.gov/disaster/coronavirus/data-resources/initial-assessment-report/key-findings-recommendations (accessed October 2, 2021).

Further Reading

1 Campbell, R. (2014). Overcoming obstacles to integrated response operations among incongruent responders. Chapter 18 in: *Handbook of Emergency Response: A Human Factors and Systems Engineering Approach* (ed. A. Badiru and L. Racz). New York: CRC Press.

3

The Challenge Ahead

There are many challenges facing crisis leaders. But the greatest challenge is ensuring that the crisis leader is ready for the next crisis. Most of this work occurs before the crisis and involves discerning the validity of prior assumptions about threats and hazards. As we explored in the previous chapters, crisis leaders are responsible for synthesizing emerging threats, trends, and vulnerabilities to clarify and define their risk profile. Crisis leaders are also responsible for engaging their stakeholders prior to an incident to address challenging organizational issues. Leaders must act on both projected risks and challenges that require external stakeholders to prevent and mitigate risks. But when the crisis occurs, crisis leaders fulfill their leadership roles in a decisive manner to ensure successful outcomes.

Learning Objectives

At the end of this chapter, you will be able to:

- Explain how threats and hazards are evolving and traditional/historical assumptions may be invalid for future planning.
- Describe the various types of vulnerable populations within a community and distinguish among their unique needs.
- Identify stakeholders within the whole community and how whole community engagement reduces the overall risk to the community.
- Identify how crisis leaders have a role in influencing external stakeholders, policies, and priorities before a disaster strikes.
- Identify six critical responsibilities of a crisis leader during and after a crisis.

3.1 Confronting the Evolving Threats, Hazards, and Vulnerabilities

There are several challenges related to crisis leadership that are governed by the previous two chapters on (1) threats and hazards, and (2) the operating context to include community-wide vulnerabilities. Threats and hazards are evolving in scope, type, and severity.

Crisis-ready Leadership: Building Resilient Organizations and Communities, First Edition. Bob Campbell, PE.
© 2023 John Wiley & Sons, Inc. Published 2023 by John Wiley & Sons, Inc.

3.1.1 Expanding Scope of Threats and Hazards

The threat and hazard list of tomorrow will be greatly expanded to include many more subdivisions of existing threats and hazards. For example, it was common for emergency managers to identify a "pandemic" as a natural hazard. The context of pandemic planning was primarily associated with influenza and various types of influenza such as H1N1 (Swine flu) or H5N1 (avian flu). With the emergence of novel coronavirus (e.g., SARS, MERS, COVID-19), the traditional context of pandemic planning is accruing new planning assumptions and variations that must be considered. Additionally, the re-emergence of infectious diseases from the past (e.g., mumps, rubella, polio) due to interconnected global mobility and a decrease in vaccination rates, will likely result in an increased frequency and resurgence of public health crises that were controlled and contained in the recent past. With the proliferation of biotechnology, many nations are experimenting with genomics and virology in new ways for defensive and offensive military purposes as well as opportunities in pharmaceutical development. In some cases (e.g., some suspect this to be the case at the Wuhan Institute of Virology), inadequate training and control measures can lead to releases into the environment and human hosts which can further transmit the disease. While not all research facilities are engaged in manipulating genes within common viruses or bacteria to increase the virility or transmission rate for nefarious purposes, this knowledge and technology is within reach and therefore warrants close observation and consideration for future threats. These specific examples highlight the emerging challenges related to "pandemic" planning, just one of many threats.

As changes in climate and environment, along with population growth impact the planet, some hazards which were rare or localized are becoming more prevalent. Leaders should be vigilant for indicators that these hazards should be added to the list. Consider areas of the US that were not prone to hurricanes and tornados, such as Pennsylvania and New Jersey. On August 29, 2021, Hurricane Ida made landfall in Louisiana. It then took a northeasterly heading through Tennessee, Kentucky, Virginia, Maryland, Pennsylvania, New Jersey, and New York. Ida's remnants and a frontal boundary caused major flooding and spawned an unprecedented number of tornados in Pennsylvania and New Jersey. This double line up of hazards resulted in confusion and deadly conditions in cases where the protective actions were in conflict. During flooding, the public is urged to stay out of low-lying areas and basements. During a tornado, the public is urged to go to low-lying areas and basements. Unfortunately, several people drowned when seeking shelter in their basement. Others, not familiar with flash flooding conditions, were swept away while driving through flooded roads. Leaders will need to consider how changing conditions are leading to hazards that were not considered in the risk profile.

3.1.2 Increasing Severity of Threats and Hazards

The severity of disasters is also increasing in many cases. Rampant wildfires throughout the western US have resulted in loss of property, life, and liberty to enjoy these communities and environments. While there are many contributing factors to the conditions that result in these severe wildfires, emergency managers and elected officials need to understand the growing threat and how this should affect public policies such as zoning, building codes, forest management, environmental protection, public awareness and notification, evacuation planning, and many more facets of governance and systems. Thankfully, these are receiving some increased attention but may require an overhaul of emergency management systems to adapt to the growing challenges. The systems of yesterday (i.e., planning assumptions based on historical statistics; obsolete building codes; public notification processes through TV, radio, and registered call-backs; reactive EOCs) need to be re-evaluated for a higher severity of threats which will require state and federal resources, area command, multi-agency coordination, and utilization of surge capacity resources.

Public officials and private organization leaders will need to monitor threats and hazards which are growing in severity so that they can project and plan for the new normal. During the summer of 2020, civil unrest spread across the US. In several cases, violent extremist organizations infiltrated peaceful demonstrations and incited crowds toward violent action. These incidents continue to grow in severity causing economic damage to communities and businesses, harm to law enforcement officials and the public, and destruction to public property. Leaders need to consider how they can prevent and properly respond to the growing intensity of this threat.

As infrastructure ages, we can expect to experience not only an increase in the frequency of infrastructure failures but also increasing severity as the contributing factors (e.g., storms, population growth) also grow in frequency and severity. Infrastructure failures may include dam failures, bridge/tunnel collapse, and transportation disruptions.

Everyone is becoming more dependent on the internet, cloud-based storage, technology, and communications networks. While these advances increase our productivity, vulnerabilities abound in cyberspace. The sophistication, frequency, and magnitude of cyber-attacks are growing as quickly as patches, malware, and virus protection are deployed. This is one area where leaders need to take a more proactive approach to preparedness as our society seems to be playing defense and catch-up after each new attack. As we consider the interconnectedness of our society and systems along with our dependency on critical operations (think aircraft relying on GPS signals, driverless cars, transportation systems, emergency communications systems, etc.), the risk and vulnerability continue to grow. Resilience demands threat detection, monitoring, redundant systems, and protections to ensure continuity of operations.

3.1.3 Vulnerable Populations

We must recognize and plan for vulnerable populations. Disasters disproportionately affect vulnerable populations. The types of vulnerabilities are as unique as the needs associated with these populations. Thus, it is worth exploring these needs and how they impact a community or organization's preparedness.

> **Vulnerability**: human induced situation caused by variables like resource distribution or availability that impacts outcomes.

There are three types of vulnerabilities: physical, psychological, and social. Physical vulnerabilities are based on physical disabilities that preclude a person from independent actions and mobility. Psychological vulnerabilities are based on differences in thought processes which may results from cultural, mental, or behavioral health issues that impact a person's ability to think, process, communicate, or act. Different cultures think about threats, hazards, and risks differently which affect how they prepare. The stressors associated with a disaster can exacerbate mental/behavioral health issues that are normally managed or under control. Social vulnerabilities are based on various demographics which may overlap. These primarily include: race/ethnicity (minority), age (elderly and minors), income (poverty), household characteristics (mobile homes, not sturdy), education (no high school diploma), language (non-English), and transportation access (no car or access to public transportation). Social vulnerabilities can affect physical and psychological vulnerabilities. While vulnerabilities are primarily attributable to individuals, there are buildings or clusters within the community of individuals who share the same vulnerability (e.g., low-income

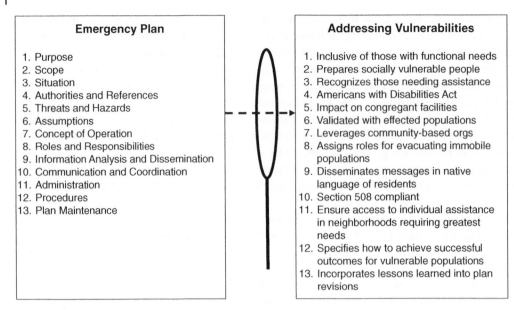

Emergency Plan	Addressing Vulnerabilities
1. Purpose	1. Inclusive of those with functional needs
2. Scope	2. Prepares socially vulnerable people
3. Situation	3. Recognizes those needing assistance
4. Authorities and References	4. Americans with Disabilities Act
5. Threats and Hazards	5. Impact on congregant facilities
6. Assumptions	6. Validated with effected populations
7. Concept of Operation	7. Leverages community-based orgs
8. Roles and Responsibilities	8. Assigns roles for evacuating immobile populations
9. Information Analysis and Dissemination	9. Disseminates messages in native language of residents
10. Communication and Coordination	10. Section 508 compliant
11. Administration	11. Ensure access to individual assistance in neighborhoods requiring greatest needs
12. Procedures	12. Specifies how to achieve successful outcomes for vulnerable populations
13. Plan Maintenance	13. Incorporates lessons learned into plan revisions

Figure 3.1 Vulnerability lens applied to an emergency plan.

housing, immigrant communities, rehabilitation facilities, etc.). Communities with high social vulnerability are typically less prepared and more likely to suffer injury or death and incur the greatest financial impact.

Leaders need to plan and care for vulnerable populations within their community or organization in different ways. Conducting a social vulnerability assessment and mapping the population against disaster risks may enable planners to proactively develop the resources and conduct the outreach needed prior to a disaster which can minimize the impact on vulnerable populations after the disaster. Knowing areas of the community that contain socially vulnerable populations can improve decision-making and resource allocation before, during, and after a disaster. Leaders must identify the vulnerable populations, anticipate their needs during a disaster, plan for the resources needed to serve and care for them, and allocate the resources necessary during and after a disaster. Since vulnerable populations suffer the greatest losses and disasters tend to widen the wealth gap, leaders have a responsibility to prevent and mitigate these impacts on their community. Figure 3.1 outlines several examples from applying a vulnerability lens to an emergency plan.

3.2 Preparing to Overcome the Challenges

3.2.1 Planning

Overcoming the challenges associated with emerging threats and hazards begins with planning. There are two major types of planning: risk-based planning and capabilities-based planning. There is certainly some overlap in the process however the outcomes are different. Table 3.1 compares the two planning frameworks.

Throughout the US, most governmental organization engage in capabilities-based planning through the THIRA process as outlined in Comprehensive Planning Guidance 201. The result of

Table 3.1 Comparison of capabilities and risk-based planning.

Factors	Capabilities-based Planning	Risk-based Planning
Risk determination	THIRA/HIRA	THIRA/HIRA
Risk-basis	Likely scenario*	Percentile-severity driven
Capability development	Based on defined scenario	Variable/total capability requirement based on risk
Result	Identification of residual risk based on defined scenario and specific threat/hazards	Gap analysis with priority action plans based on reducing total risk to community
Mission areas	Response and recovery focused	Prevention, mitigation, protection, response, recovery
Risk management	Funding projects based on type of grants (restricted based on grant limitations (e.g., no equipment), increased capabilities by function (e.g., fire or police department))	Funding prioritized across capabilities to reduce overall risk profile. Vulnerabilities may be addressed through whole community integration of resources

*Some communities conduct a worst-case analysis which does not typically enter into the calculus for developing capabilities due to the overwhelming severity of the situation.

capabilities-based planning is the identification and development of capabilities (defined by a set of resources – human, equipment, knowledge, etc.) to respond to a defined threat or hazard. This begins with identifying hazards and threats, then conducting an analysis based on likely (most probable) scenarios. In some cases, communities will decide to rely on external support to deliver the capabilities needed to respond effectively. But in other cases, communities will build the capabilities needed to respond to the incident of concern. This type of planning is focused on response and recovery.

Risk-based planning begins the same way, acknowledging that a community cannot plan for every level of severity for each threat, but creating a risk profile across multiple threats and hazards. Risk-based planning enables the planner to define the threshold of unacceptable risk (e.g., greater than the 80th percentile of severity), weigh the threats and hazards according to their respective probabilities of occurrence, and compute the risk in terms of consequences (e.g., casualties, property damage, economic impact, etc.). Based on the overall risk profile and adverse impacts, the planner develops capabilities to control and mitigate the risk. In this type of planning, risk ultimately drives the capability requirements and forces a conscious decision over the level of risk tolerance the community is willing to accept. The key difference is the means of determining the capability requirement (top-down based on risk or bottom-up based on capabilities) and resulting actions.

3.2.2 Funding Systems

To overcome the antiquated capabilities-based planning system, changes are needed in how federal grants are delineated, funded, and prioritized. A systems approach to prioritizing and allocating funds based on total risk and a community's leveraged actions (e.g., changing flawed zoning laws that permit people from building houses in flood-prone locations since the federal flood insurance and mitigation funding underwrites these losses). The Disaster Recovery Reform Act takes several bold steps in this direction. Some communities pool their grant funding and allocate it based on need, rather than the best grant writer in the county or department. However, when mitigation plans neglect hazards due to an over-emphasis on flooding, or grants deny equipment purchases or sustainment because the grant can only be used for planning or training, emergency managers are left trying to piece together a mosaic of funding to improve their community preparedness driven by available grants.

3.2.3 Organizational Approach

The operating context and traditional assumptions need to be reconsidered as well. Our response to most crises involves a heavy component of government which is charged with protecting public health and safety. Emergencies begin at the local level, which is where preparedness begins, mitigation measures are implemented, and resilience is determined. When additional response resources are needed due to the scope and scale of the emergency, the locality requests support from the state. States are ultimately responsible for managing a disaster within their state, particularly when the disaster crosses local boundaries and requires additional resources. The state does not direct responses in local communities, but rather manages state-wide resources and provides these resources to the local community as requested. Due our federalist system of government, the federal government's role is limited to supporting states upon request. The federal government does not go into a state or local community, uninvited, and direct or manage the response (except in cases of terrorism, foreign interference, etc. – where the federal government has jurisdiction). This construct is the basis of the National Frameworks: locally executed, state managed, federally supported.

But also consider that the private sector owns 85% of the critical infrastructure which is necessary for normal operation within our society. The private sector bares a significant onus for the systems and infrastructure that the public rely on to live, work, and play. While large utilities, transportation systems, heavy industry, and health care are engaged in preparedness activities, some businesses neglect their role in preparedness, perhaps not valuing their role in the supply chain in the same way that the end consumer values the end-product. Recent disruptions in logistics have many customers and businesses wondering why there appears to be a shortage of truck drivers which is impacting the timeliness of product deliveries to include inputs into manufacturing end products. Other institutions such as schools, churches, charitable organizations, and community-based organizations are also recognizing their importance to society. Their lack of preparedness in many cases is causing dramatic shifts as some institutions are shuttered while those that were prepared are thriving.

Finally, the public must own the culture of preparedness as well. As we have watched many systems, businesses, and institutions fail during the pandemic, we have learned an important lesson in the fragility of our society and how cascading impacts from disruptions can grow and affect our lives. While we cannot reduce the risk to zero, our preparedness decisions determined several outcomes related to our ability to survive or thrive during the pandemic.

3.2.4 Rethinking the Norms

We need to continually re-examine the traditional approaches to disasters. Former FEMA Administrator Brock Long recently proposed a new approach to hurricane preparedness. He proposed that we stop trying to evacuate areas in the hours prior to a hurricane making landfall and instead invest in infrastructure to withstand the hurricane winds and associated flooding. Unfortunately, many communities lack the road capacity needed to evacuate the population in a timely manner. Instead of long-distance evacuation which is also very difficult for economically challenged persons that lack the financial resources to afford housing in another location, or perhaps do not own reliable transportation to get them to a safe area, Brock Long proposed establishing community shelters which could accommodate people without evacuation to distant places. Novel approaches such as this are needed as we rethink the operating context in light of tomorrow's threat profile.

3.2.5 Whole Community Preparedness

There is also a role for the private sector. US FEMA has outlined the concept of "whole community preparedness" in its national preparedness goal. True whole community preparedness fully engages citizens, businesses, government, and non-profits to collaboratively prepare for incidents that may impact the community. Closer coordination and trust are needed among community members, entities within each community, and the public sector emergency managers, responders, and stakeholders to enhance preparedness, facilitate response, and accelerate recovery. This need was most recently illustrated by the impact of COVID-19 on the whole community. The lack of coordination between employers, school boards, and public health officials has hampered economic recovery, adversely impacted learning, and has pitted public health officials against businesses and schools, creating binary choices between health and education, and health and economic recovery. Whole community planning brings together all stakeholders to express their respective concerns and interests so that a collaborative course of action can be developed to address the community's interests as a whole. Instead, our communities received many disparate approaches with different standards for different groups. For example, our school district decided to delay in-person learning because a significant number of their teachers lived in another county, and if that school district did not reopen, our local school board surmised that it could not staff the schools. This is something multi-jurisdiction coordination could have solved with a unified decision. A different approach is needed, and FEMA is on the right path with this important element of the nation preparedness goal.

We are going to become increasingly challenged by emerging threats and increasing severity of disasters which will require more operational coordination among the whole of community and all levels of government in ways that are only occasionally executed or exercised. Many of the preparedness systems that we have in place are not flexible and robust enough for tomorrow's responses. Our communities lack the resiliency needed due to systemic issues related to funding lines, local governance decisions, lack of strategic economic development initiatives to diversify and build resilience into local economies, lack of infrastructure initiatives to maintain pace with population growth, utilization, and decay, and obsolescence.

Leaders and decision makers must engage in making incremental improvements to our systems to build resilience before the disaster strikes. This book could outline all of the great ways to lead through a crisis when it hits, but the outcome will not be decided by the leader during the crisis. The outcome will be decided by great leadership *before* the crisis.

3.3 Creating Conditions for Success before the Crisis

Making sound decisions to positively affect outcomes begins before disaster strikes. Many of these decisions are made outside of your functional responsibility in your community or organization. Sometimes these decisions are made at the individual, constituent level. Still some decisions are made at the federal or international level. Regardless, a leader's role is to lead through influence to set the conditions for success. Unfortunately, this is the hard, thankless work that is often met with resistance, competing interests, and bureaucracy. We will explore a few examples to provide some context.

Example 1: Approximately 35% of communities have adopted the latest building codes. Building codes are the minimum standard to which buildings must be constructed. Many building codes were based on obsolete data related to flooding, winds, earthquake probability, and availability of building materials. After watching or reading about the "Last House Standing" in Mexico Beach, Florida, it becomes obvious how ineffective modern building codes are against modern hazards. This example also illustrates the individual responsibility to build or buy a home that is constructed to withstand a disaster. So, what is your role in ensuring that your community and home are built to an adequate standard to withstand the risks facing your community?

Example 2: Local zoning boards control the designation of areas within a community as residential, industrial, agricultural, or other designations. As communities grow, inadequate urban planning results in conflicts among space use. A desire for more property tax revenues and growing the industrial base results in construction in risky areas such as flood plains. Residential homes that sprawl into previous industrial zones increase nuisance hazards such as noise, odors, and increased heavy vehicle traffic. As the city of West, Texas grew, houses, a school, and a nursing home were built in close proximity to the West Fertilizer Company. On April 17th of 2013, this facility caught fire and exploded. This incident resulted in the injury of more than 200 citizens and the death of 15 people. What is the role of a leader in providing input to the zoning board?

Example 3: There are numerous federal grants that flood into communities for specific purposes: purchasing firefighting equipment, hazardous materials training, pre-disaster mitigation and resilience, emergency planning, catastrophic planning, and the list goes on. How should organizations and grant recipients coordinate on spending these funds to increase the right capabilities needed to close resource gaps when the funds and recipients may not align with the greatest risks to the community?

Example 4: Eighty-five percent (85%) of the critical infrastructure in the US is controlled by the private sector. These companies strive to remain competitive, productive, and profitable to ensure their continuation while making measured investments in future technology, capital improvements, and growth. What obligations do these private sector companies have to the customers they serve? Should they raise prices so that they can invest more in preparedness and resilient infrastructure? What role does the public sector have to collaborate, share information, and support these companies before, during, and after a disaster? Should the public sector fund some portion of the private sector's preparedness activities to enhance community resilience?

As illustrated through these examples, pre-disaster engagement outside of your immediate organization can determine preparedness and resilience strategies which have an impact on the consequences of a disaster (i.e., its duration and impact across the community). While you may take the actions needed for your department or organization to be prepared and invest in resiliency, you are dependent on others outside your organization to make the right decisions too.

Building a culture of preparedness requires a whole community approach to engage those that could be impacted. It requires a network of interested parties who are aware of the threats and hazards that they face within their organization, in their community, among their customers and constituents, and throughout their supply chain. Crisis leaders help these constituencies recognize the potential problems they could face during a disaster so that they can lead them through a coordinated pathway to preparedness and resiliency before the disaster. Later in this book, we will explore the role of Community Lifelines in pre-incident planning, stabilization in response, and restoration in recovery. Identifying vulnerabilities can lead to meaningful planning and decisions to prevent and mitigate the impact of a disaster. The decisions and actions taken prior to a disaster can prevent loss of life, injuries, and financial distress. Appropriate mitigation measures can accelerate recovery with less impact from the hazard. FEMA's Building Resilient Infrastructure and Communities (BRIC) program is based on the premise that reallocating money from disaster recovery to mitigation and resilience will result in a return in savings. What investments do you need to make now in order to prevent excessive losses in the wake of a disaster?

3.4 Leading through a Crisis

Leading through a crisis begins with building the right systems and resilience *before* the crisis. It is better to face a crisis with a whole community that is resilient and prepared, than a community that is not ready. We will have further topics and discussions on preparing for the crisis and establishing resilient organizations and communities, but in order to fully address the challenge ahead, we need to establish a foundation for leading during a crisis. Once the crisis begins, a leader is responsible for paving the way through the crisis as safely and quickly as possible. Every leader needs a playbook to guide them through a crisis and should possess some situational awareness of their capabilities, risks, and vulnerabilities. Some of the key steps in leading through a crisis are outlined in the following paragraphs.

3.4.1 Establish the Goal and Objectives

Everyone needs direction and a leader provides the ultimate course to resolution by setting a goal. If those involved in the response do not know the goal, everyone will do what they think is best (which will not align with the overall effort). It may be easy for each organization involved in the response to get caught up in doing what they know best (i.e., they may view the problem and solution set through the lens of their discipline rather than the broader, multidiscipline perspective). Alignment and unity of effort is critical. Incident commanders and leaders must establish incident objectives as part of the incident action plan (IAP) early in the response. The IAP drives the response priorities and actions. "Where there is no vision, people perish." [1] Responders need to know the leader's intent and the end state so that they can align their capabilities with the requested activities and tasks to achieve the goal. One additional tool covered in Chapter 7 is the Community Lifeline toolkit which creates a framework for stabilizing and restoring essential services within a community. Ultimately, a leader's first responsibility is to establish goals and objectives.

3.4.2 Communicate the Goal and Incident Objectives

Everyone needs to know the goal and objectives. An incident commander will typically use the ICS structure to communicate information but a major disaster with multiagency, multijurisdictional response, a more robust communication strategy is needed. Each team needs to know their role, what task they need to complete, when they need to do it, and how they support achieving the goal and objectives. As incident objectives are accomplished and new objectives emerge, the leader needs to keep the team of responders informed and involved. Additionally, coordination centers will be monitoring progress and anticipating needs. Leaders need to share information to keep the team informed of progress, successes, failures, and next steps. In major incidents, there will likely be additional stakeholders such as the public, media, businesses, volunteer organizations, and other governmental organizations. Leaders need to stay ahead of the information cycle by establishing a briefing schedule to align internal and external stakeholders. This could include leveraging the public information officer and the liaison officer.

3.4.3 Assign Resources

Incident objectives require resources to execute – human, infrastructure, information systems, equipment, and financial. Section chiefs, business leaders, elected officials, and volunteers need resources to do what is necessary. In some cases, these resources are available but in other cases, these may be requested through the Emergency Operations Center, elected officials, businesses, and volunteers. Typically, the Logistics Section chief is responsible for gathering and providing resources to the Operations Section. Identifying, mobilizing, deploying, and employing resources can be a time-intensive activity requiring patience, persistence, and effective communication. Pre-incident planning may pre-identify the resources needed to stabilize and restore essential services and functions. But during a crisis, leaders will need to leverage their networks and relationships to efficiently access necessary resources.

3.4.4 Coordinate with Stakeholders

Leaders will encounter various constituents who may not be direct reports or involved in response. However, leaders can expand their relational capital and enhance their situational awareness through stakeholder engagement. Listening, understanding, and empathizing with stakeholders builds relational capital and may leader to productive collaboration. Leaders must understand stakeholder perspectives in order to build trust and secure cooperation when needed. Examples of these stakeholders may include the media, adjacent landowners, nearby agricultural industry, those affected by water or land use, concerned public, political leaders, and community-based organizations. If they are not working with you, they may be working against you. Therefore, take the time to build trust, listen to enhance buy-in, and engage to build accountability.

3.4.5 Monitor Progress and Control Outcomes

Leaders enhance their situational awareness and decision-making by monitoring progress in relation to incident objectives. This may also include intermediate objectives and specific lines of effort. Leaders should establish a formal and routine reporting process to ensure information is updated, reported, and shared as needed with those involved. Monitoring and measuring progress creates a moment of truth regarding the effectiveness of the actions taken. This creates a decision point: continue with the same activities or make an adjustment. The outcome is controlled through the decision-making process. If the desired progress toward the expected outcomes is not being achieved, leaders must recognize and adjust their actions. This may include modifying the objectives, resource allocation, or operating conditions. It is important to remember that each decision takes time to implement; this time-to-implement must factor into the decision-making process.

3.4.6 Communicate Again

Communication is frequently cited as a problem in after-action reports. Leaders need to exercise clear and regular communication in order to share relevant information with responders and external organizations that may be affected. Do not underestimate the importance of regular, frequent, and repetitive communication.

These keys to crisis leadership strike a balance between management by objectives and relational influence. Chapter 11 will dive deeper into the essential attributes needed to accomplish this. While these steps may prove challenging without adequate preparedness and practice, a crisis-ready leader needs to follow a system or risk becoming part of the crisis.

Summary of Key Points

- Threats and hazards are always changing in scope, type, and severity. As disasters continue to evolve and worsen, leaders must continually revise their risk profile and account for emerging threats and hazards in order to project and prepare for future incidents.
- Risk-based planning is essential, not because it can account for every possible event, but because it provides a realistic understanding of the variations in hazard occurrence and severity so that leaders can adapt and allocate the right resources to prevent, mitigate, and respond to incidents.
- Whole community engagement is essential when creating risk-based plans. This type of engagement draws in all stakeholders such as citizens, businesses, government, and non-profits (not just responders) so that all facets of a community can engage in preparedness.
- Crisis leaders can have the biggest impact prior to the crisis by setting goals and making decisions that will create a resilient community. This requires anticipation and recognition of hazards, the risk profile, how quickly the risk profile is changing, and vulnerabilities which need to be addressed. Influential leaders build a coalition among external parties to collaboratively address these challenges.

Keys to Resilience

- Crisis-ready leaders are engaged in preparedness and resiliency long before the crisis occurs. Leaders work to influence external parties, policy makers, local officials, boards, and partners to make them aware of risks and implement effective counter-measures that enhance resiliency.
- Leaders coordinate and collaborate with the whole community of partners to define roles, responsibilities, resources, and organizational interfaces. These organizational and individual relationships are key to rapid intervention, response, and recovery operations.
- Leaders are prepared for the next crisis because they engage their team in planning for emerging risks.
- Leaders understand their changing risk profile and have methods for detecting risk acceleration and acting decisively to address increasing risks in a timely manner.

Application

1) Consider the threats and hazards facing your organization or community. Prioritize them based on risk. According to most emergency management standards, organizations should start with a risk assessment of their threats and hazards [2, 3].

2) Identify emerging threats or hazards that need to be added to your organization's Hazard Identification and Risk Assessment (e.g., what events have been in national or international news over the past year that are not part of your hazard risk profile)?

3) Compare the severity of existing threats or hazards over time. Which threats or hazards have become more severe in your community? What preparedness measures need to change in order to adequately prevent, mitigate, respond, and recover from these more severe hazards?

4) Review the demographics of your community or organization. Identify vulnerable populations. What are their access or functional needs? How are you addressing those needs in your plans and preparedness program? How are you engaging with them to include them in creating a culture of preparedness?

5) What external organizations are you currently engaged in coordinating or collaborating to strengthen preparedness? After reviewing the examples in this chapter, what other organizations should you engage to enhance preparedness and resiliency?

6) Crisis leadership begins with solid leadership practices during routine operations. Self-assess how you are doing with implementing the six steps of navigating through a crisis in routine, non-crisis situations. Rank them from strongest to weakest and identify actionable steps to improve weaknesses as these weaknesses will become more pronounced during a crisis.

7) Reflecting on your organization or community's capabilities and vulnerabilities that would prevent you from smoothly implementing these steps: What systems, plans, infrastructure, equipment, personnel, or resources do you need to enhance? Keep in mind that the total cost of response and recovery generally increases with the duration of the incident. What actions can you take to ensure swift execution of response actions that will lead rapidly to recovery?

References

1 The Holy Bible: King James Version (1769). Cambridge Edition. King James Bible Online (2022). Proverbs 29: 18. www.kingjamesbibleonline.org (accessed May15, 2022).

2 ANSI/EMAP EMS 5-2019 (2019). *Emergency Management Standard.* Falls Church: Emergency Management Accreditation Program. www.emap.org (accessed October 2, 2021).

3 NFPA 1600 (2019). *Standard on Continuity, Emergency, and Crisis Management.* Quincy: National Fire Protection Association. www.nfpa.org (accessed October 2, 2021).

Further Reading

1 Federal Emergency Management Agency (2011). A whole community approach to emergency management: principles, themes, and pathways for action. https://www.fema.gov/sites/default/files/2020-07/whole_community_dec2011__2.pdf (accessed October 2, 2021).

2 Federal Emergency Management Agency (2018). Threat and hazard identification and risk assessment and stakeholder preparedness review guide: comprehensive planning guide 201. 3rd Edition. https://www.fema.gov/sites/default/files/2020-04/CPG201Final20180525.pdf (accessed October 2, 2021).

3 Lencioni, P. (2012). *The Advantage: Why Organizational Health Trumps Everything Else in Business.* Hoboken: Wiley.

4

Risk-Based Planning

In order to develop relevant plans, programs, and preparedness initiatives that adequately address an organization's risks, leaders need to know their risk profile. Without a risk profile, decision-making will tend to focus on the most recent incident, most likely hazards, or the most severe threats which have dominated the news. This chapter will provide specific information on how to conduct a hazard identification and risk assessment so that leaders can allocate resources and planners can make relevant assumptions and investments based on risk.

Learning Objectives

- Identify resources and methods for identifying threats and hazards.
- Perform a risk assessment applying the methodology outlined in this chapter.

4.1 Hazard Identification Risk Assessment Methodology

While there are many methods [1] for conducting a risk assessment, we are going to focus on a basic method that aligns with the Emergency Management Standard (ANSI/EMAP EMS 5-2019), a consensus standard used to accredit emergency management programs [2]. The benefits of conducting a Hazard Identification and Risk Assessment (HIRA) include:

- Improved risk communication by establishing a common language.
- Setting data-driven planning priorities based on risk.
- Justifying planning priorities.
- Evaluating and comparing prevention, protection, and mitigation actions.
- Anticipating challenges during an incident.
- Whole community engagement among the public sector, private sector, and the public.

While the risk assessment results in a rank-ordering of hazards based on risk, it also identifies hazards that could influence the risk of other hazards (e.g., prolonged winter storm or flooding could cause water disruptions; earthquake that causes infrastructure failure; flooding that results in hazardous material release into the environment). Other environmental conditions such as sea-level rise and land subsidence over time may increase the risk of hazards such as flooding, shoreline erosion, and tropical coastal storms. Finally, during recent years, we have seen an increasing

Crisis-ready Leadership: Building Resilient Organizations and Communities, First Edition. Bob Campbell, PE.
© 2023 John Wiley & Sons, Inc. Published 2023 by John Wiley & Sons, Inc.

frequency and severity associated with civil unrest, active threats, and cyberattacks while enduring a pandemic. This highlights the need to plan for multiple simultaneous threats, changing risk profile, and synergistic and complicating effects on response and recovery operations.

It is expected that the HIRA will be used by leaders to guide preparedness activities (e.g., planning, training, exercises) and decision-making (e.g., resource allocation). The planning process begins with identifying hazards and risks associated with those hazards. Planners and decision-makers should prioritize higher risk hazards over lower risk hazards, rather than focusing on only one dimension of the risk assessment (i.e., likelihood of occurrence or consequences such as severity of impact). A risk-based approach to planning and resourcing ensures capabilities are developed in a manner that optimally reduces overall risk to the community or organization considering limited resources. The method provides a scientifically defensible foundation for preparedness decision-making.

4.2 Risk Calculation

Risk is most easily understood by analyzing the component parts of risk. The HIRA defines the components of risk in the following manner: **Risk** is a function of **Likelihood** and **Consequence**. **Likelihood** is determined first by calculating the historical occurrence and then estimating the future likelihood based on trends and subject-matter expert-projections. **Consequence** is determined based on **Warning Time** prior to the incident, **Duration** of the incident, and **Impact** of the incident across multiple domains. These domains include impacts to property, health, and safety, critical facilities, response capacity, the environment, the economy, and standard of living/quality of life. Figure 4.1 outlines this relationship visually:

It is common to see risk calculated as the sum or product of likelihood and consequence (e.g., *LIKELIHOOD + CONSEQUENCE, or LIKELIHOOD * CONSEQUENCE*) or a function of threats, assets, vulnerabilities, and/or consequences (e.g., *RISK = THREAT * VULNERABILITY * CONSEQUENCE*). The latter formula is similar to the FEMA National Risk Index which is an excellent and accessible resource for determining community-wide risk from natural disasters: *RISK = EXPECTED ANNUAL LOSS * SOCIAL VULNERABILITY/COMMUNITY RESILIENCE.*

There is some overlap among these approaches, but we will defer to the first approach to align with the Emergency Management Standard and ensure that we review an unmitigated view of the relative risks posed by hazards. This approach also allows for adaptability and normalization across unquantified and emergent hazards or threats. The second formula is commonly used in the security discipline to account for the vulnerability of a target in the risk assessment. The final formula is appropriate when each of the variables can be quantified on a standardized basis such as when considering natural disasters, but not man-made or technological hazards.

To collect data and calculate risk ratings, a risk calculation tool may be useful. A risk calculation tool converts hazard information into a set of numerical scores that allow for comparison across many natural and manmade hazard types. With the appropriate information, any hazard can be processed through the risk tool for comparison with HIRA data. Similarly, any existing HIRA risk score can be easily interpreted by referencing the quantitative levels outlined in the risk tool. The tool's design ensures that the resulting risk score is transparent, replicable, and scientifically defensible. By ensuring a consistent approach to developing the input scores, the resulting relative risk scores are comparable.

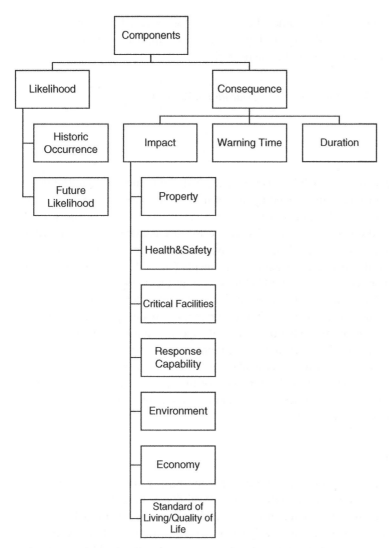

Figure 4.1 Risk components.

Each hazard is assigned a numerical score in each of the following four risk assessment input categories: likelihood, impact, warning time, and duration. Numerical scores may range from one to four based on criteria that are defined explicitly in the tool. The scores from each section are multiplied by an assigned weighting factor as determined by the planning team. For example, likelihood may be weighted at 50% of the Risk Score. Consequence may be comprised of impact (40%), warning time (5%), and duration (5%) for a combined total of 50% of the risk score. Once multiplied by the weighting factor, the sum or product of the scores becomes the total Risk Score for the hazard. Using a product, instead of the sum, stretches the distribution of risk scores over a larger scale and accentuates the effect of higher input scores on the overall risk score. This method may be helpful in stratifying high, medium, and low risk hazards. Using the sum of the input scores for each hazard as shown in the equation below treats the distance between each input rating as equal (e.g., the incremental difference between ratings of one and two is the same incremental effect as between three and four). When deciding whether to choose between these approaches, it is important to consider the criteria used in the rating system. Either approach can be acceptable as long as there is a consistent approach to assigning the ratings.

$$Risk = \left(Likelihood \times 50\%\right) + \left(Warning\, time \times 5\%\right) + \left(Duration \times 5\%\right) + \left(40\% \times \sum_{i=1}^{n} \frac{Impact\, rating_i}{n}\right)$$

4.2.1 Likelihood Rating

The likelihood rating is based on the probability of occurrence. This data can be collected based on a jurisdictions historical occurrence data divided by the number of years researched in order to determine an annual occurrence rate. Natural disaster historical data is generally available through NOAA and USGS websites. Alternatively, natural disaster occurrence data has been compiled in the FEMA National Risk Index. Man-made data may be available in other sources, but we have found that a tool used for water risk and resilience assessments (i.e., EPA's Vulnerability Self-Assessment Tool) has incorporated projections for likelihood of some technological and man-made hazards. Nonetheless, exact details may not be necessary. Cross-discipline team public safety professionals may provide qualitative data related to their experiences responding to these incidents in recent years. Planners involved in reviewing this data and collecting these experiences should seek to estimate the future likelihood of occurrence given emerging trends, weather cycles, and other factors. To simplify and standardize the process, likelihood may be represented using a Likert scale: *Unlikely, Infrequent, Likely, Very Likely*. Each descriptor may be associated with ranges of annual probability (e.g., Unlikely represents less than 1% annual probability) and assigned a rating between one (Unlikely) and four (Very Likely).

4.2.2 Consequence Rating

A similar process could be followed for warning time and duration. Warning time represents how much time exists before the incident. With some natural disasters such as a hurricane, warning time may be days. With unexpected man-made incidents such as cyberattacks, there is usually no warning time. Again, planners may assign a one to four rating based on the amount of warning time (e.g., one may indicate very long time such as over 24 hours, and four may indicate a short time such as less than six hours). Duration of the incident could be quantified as short (i.e., less than six hours) to very long (greater than one week) on a one to four scale.

Finally, planners may assess the impact of each hazard on multiple domains such as the public, responders, continuity of operations, property, facilities, infrastructure, environment, economic condition, and public/stakeholder confidence. A team of experts representing each of these domains can assess the level of impact on a one to four scale, where a rating of one is "limited" and four is "catastrophic." The average of all of these ratings becomes the impact rating which in our example is weighted at 40%. Ultimately, impact, duration, and warning time represent the consequences for the hazard. When combined with the weighted rating for likelihood, then the risk rating cam be computed.

After gathering this data from stakeholders and experts within the community and external stakeholders (e.g., NOAA, FEMA, trade groups, etc.), planners can calculate a risk rating for each threat and hazard so that they can be compared with each other. Figure 4.2 illustrates an example risk profile for the civil unrest hazard.

Civil Unrest Risk Profile				
	Risk Assessment Category	**Likely Hazard Scenario**	**Worst-Case Hazard Scenario**	**Weight**
Likelihood	**Likelihood**	3 Likely	2 Infrequent	50%
Consequence	**Impact**	1.50 Limited-Significant	2.50 Significant-Critical	40%
	Warning Time	4 Short	1 Very Long	5%
	Duration	1 Short	3 Long	5%
Total Risk Score		2.35	2.20	

Figure 4.2 Example risk profile for civil unrest.

4.3 Vulnerability Assessment

Using a standardized method yields comparable risk ratings to enable prioritization of resources based on risk. Planners should also consider their organization or community's vulnerability across the rank-ordered list of risks. Vulnerability assessments can be subjective, but it is important to consider the vulnerability associated with each risk so that leaders can identify and close any gap by building capabilities and capacity. In our model, risk is based on external threats and hazards to the organization or community. But vulnerability considers the capabilities (or the lack of capabilities) to prevent, mitigate, or respond to each risk.

There are several approaches to conducting vulnerability assessments. A common approach is the POETE model which considers Plans, Organization, Equipment, Training, and Exercises. My preference based on 18 years of experience conducting vulnerability assessments is to use the DOTMLPF framework which is an acronym for Doctrine, Organization, Training, Material, Leadership, Personnel, and Facilities. Under each model, analysts explore the capabilities based on various resources, competencies, and categories to identify gaps. Keep in mind that each risk needs to be examined across all disciplines and mission areas of emergency management (i.e., prevention, protection, mitigation, response, and recovery). It is also worth reviewing after action reports from exercises and real-world events to ensure that each of these components can been verified and validated. The outline below lists some of the key questions or criteria for consideration when applying the DOTMLPF model. Here is an example of how this works.

Doctrine: How well do plans, policies, and procedures adequately address this risk?

Organization: Does the organization have an incident management system established to interface with internal departments and external entities as needed to address this risk?

Training: Have those responsible for preparing for this risk been trained to the appropriate level? Do they possess the knowledge, skills, and abilities to perform their tasks before, during, and after the incident with competence and proficiency?

Material: Do responders have the right equipment, supplies, and accessories needed to adequately address this risk? We often think about response equipment, but also consider warning and notification systems such as flood and stream gauges, air quality sensors, and information system protection.

Leadership: Is the leadership team crisis-ready? Are they qualified, trained, and prepared to lead through a crisis? This book explores this topic in more detail. Are leaders formally appointed or assigned to positions during a crisis? Does the leadership team have experience with this risk?

Personnel: Is the organization adequately staffed to prepare for and respond to this risk? The answer to this question is usually "no" but consider what it would take across each discipline to adequately prepare and respond. Are mutual-aid agreements, memorandum of understanding, or other agreements in place to access and leverage personnel when needed for surge capacity? What is the estimated time of arrival for receiving these teams after the request?

Facilities: Does the organization have adequate infrastructure to address this risk? Consider facility design standards, zoning, and ratings for wind, weight capacity for the roof, and occupancy capacity and accessibility.

There are many more questions that could be asked under each category. But I think you will find that as you start asking and responding to these questions, more questions will naturally follow which will lead to uncovering vulnerabilities. Like our risk rating exercise, each vulnerability can be rated based on the extent that it causes vulnerability for the organization. For example, no impact, limited impact, significant impact, critical impact, or catastrophic impact. Impact can be qualified in terms of organizational downtime, lost revenue, personnel impacted, and so on.

Leaders can use vulnerability ratings in conjunction with risk ratings to reprioritize preparedness activities and align initiatives and resources. One approach to integrating both of these aspects is to multiply the risk rating and vulnerability rating for each hazard, then rank-order the resulting list from highest to lowest. Whichever methods are used, leaders need to understand risk and vulnerability so that they can prioritize resources, inform decision-making, prepare for relevant hazards, and build resilience before the incident occurs.

Summary of Key Points

- Risk assessments enable comparison and prioritization of threats and hazards.
- Risk is based on likelihood and consequences, although there are other approaches which incorporate additional factors.
- Impact analyses are most credible when the right stakeholders and experts are engaged in the process.
- Vulnerability assessments expose internal weaknesses and deficient capabilities that are necessary to adequately address the risk posed by each hazard.
- Leaders should prioritize based on risk, not frequency of occurrence or possible consequences.
- The FEMA National Risk Index is a great starting point for gathering historical data for natural disasters.

Keys to Resilience

- Resilience is a mitigating factor to risk and vulnerability.
- Leaders prioritize resilience initiatives that strengthen their organization or community in preparation for the next incident.
- Conduct risk and vulnerability assessments to inform resilience initiatives.

Application

1) Does your organization or community have a HIRA? What are the top risks? How is it inform-ing leadership priorities, decisions, and resilience initiatives?
2) If your organization does not have a HIRA, assemble a cross-disciplinary team and work through the process outlined in this chapter.
3) What are some of the impacts that high-risk hazards could have on your organization or community? What are some potential initiatives to mitigate the impact and recovery time?

References

1 ISO/IEC 31010:2019 (2019). *Risk Management–Risk Assessment Techniques*. Geneva: International Organization for Standardization.

2 ANSI/EMAP EMS 5-2019 (2019). *Emergency Management Standard*. Falls Church: Emergency Management Accreditation Program. www.emap.org (accessed October 2, 2021).

Further Reading

1 Aven, T. (2016). Risk assessment and risk management: review of recent advances on their foundation. *European Journal of Operational Research* 253 (1): 1–13. doi: 10.1016/j.ejor.2015.12.023.

2 Federal Emergency Management Agency (2021). National risk index. https://hazards.fema.gov/nri (accessed October 2, 2021).

3 Ning, M. and Yijun, L. (2020). Risk factors and risk level assessment: forty thousand emergencies over the past decade in China. *Journal of Disaster Risk Studies* 12 (1): 916. doi: 10.4102/jamba. v12i1.916.

4 U.S. Department of Homeland Security (2022). Risk assessment. https://www.ready.gov/risk-assessment (accessed May 15, 2022).

Part Two

Decision-Making

A high degree of situational awareness can improve decision quality. Aligning the right decisions creates a line of effort that can achieve a desired outcome. In chapter 5, we will explore Dr. Endsley's model of situational awareness along with internal and external factors that can impact situational awareness. By understanding how to improve situational awareness, leaders can optimize their organizational structure, communications, tools, and information management systems. In chapter 6, we will explore decision theory to equip leaders with several frameworks for decision-making. Finally, in chapter 7, we will provide information on how to apply the Community Lifelines toolkit from a macro-perspective during major incidents that require stabilization and restoration of the seven lifelines.

5

Situational Awareness

Navigating a crisis begins with understanding the circumstances, defining the problem to be solved, and establishing decision points along the pathway of response to recovery. This chapter will focus on understanding the circumstances through situational awareness.

Learning Objectives

After reading this chapter, you will be able to:

- Understand Dr. Endsley's model of situational awareness.
- Identify internal and external factors that impact situational awareness.
- Synthesize available information in order to create situational awareness.
- Evaluate decisions that were made during various incidents with regards to situational awareness.

5.1 What Is Situational Awareness?

> **Situational awareness** refers to the degree of accuracy by which one's perception of his current environment mirrors reality [1].

The US Coast Guard defines it simply as "Knowing what is going on around you." (USCG, 1998) In any situation, a leader has a perception of the situation that is influenced by what information has been shared, their experiences from similar situations, and observations. Unfortunately, "everything" is not immediately known. Especially in the early stages of the situation, only pieces of the puzzle are available. As time progresses, additional pieces of the puzzle are made available, but it takes someone with knowledge, skill, and insight to find where that piece fits in the puzzle in order to bring more clarity and improve situational awareness.

First, the leader needs value-added pipelines of data flowing into a fusion center where an expert can make sense of the data, thereby converting it into useful information. Second, this information must be useful in contributing to a decision that needs to be made in order to progress through the incident response toward recovery. It is important to differentiate useless data from useful data; therefore, it is not about how much volume of data or information can flow to the decision maker. Too often, decision makers are overwhelmed with data (some of it useful) which inhibits them

Crisis-ready Leadership: Building Resilient Organizations and Communities, First Edition. Bob Campbell, PE.
© 2023 John Wiley & Sons, Inc. Published 2023 by John Wiley & Sons, Inc.

from viewing the useful information and formulating good decisions. The objective is to increase the usefulness of information in operations through the consistent application of polices, exchange standards, and common frameworks. This creates conditions for sharing the right information with the right people at the right time. Technology can be helpful in doing this, but sometimes, technology just increases the volume of data available to decision makers [2]. Throughout the rest of this chapter, we will explore a situational awareness model and its relationship with decision-making.

5.2 Situational Awareness Model

The situational awareness model developed by Dr. Mica Endsley highlights the hierarchy of information based on three levels of situational awareness [3]. Each level is described in detail in the following sections. An additional level of intervention was added by Burdick to highlight the role of intervention in changing the outcome. This is depicted in Figure 5.1.

5.2.1 Perception

Level 1 is the Perception of the elements in the environment. This involves recognizing relevant status, attributes, and dynamics in the environment. An example of this would be seeing dots on a map of varying colors and knowing what each of these colors represents. Perception can be described as receiving data from an incident. These are raw facts without context and are limited by what people perceive, which influences their perception of reality.

During the recent COVID-19 pandemic, perceptions included number of cases, hospitalizations and deaths, and available resources such as hospital beds and ventilators. But this raw data was meaningless without the context of population, community centers, age distribution of the population, degree of community spread, and projections. Data gathering is essential, but it is also important to gather the right data that is relevant to the incident encountered.

During the West Fertilizer Company incident in 2013, some of the data available to first responders included the fact that there was a fire at the fertilizer company, the color of the flames, and the presence of anhydrous ammonia tanks at the facility. Missing data relevant to decision-making included: distance to the nearest fire hydrant, capacity of water, knowledge of the potential for detonation of the fertilizer, Hazmat operations level training, and evacuation area around the

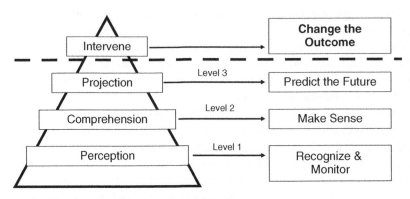

Figure 5.1 Situational awareness model [3, 4].

facility. Unfortunately, this incident resulted in the death of 12 responders and 3 members of the public, with over 200 hospitalizations. Good situational awareness starts with preparedness and could include pre-incident plans that show the nearest source of water for fire suppression, evacuation radius, chemicals stored on-site along with safety data sheets, and initial incident action plans devised precrisis.

Perception is the sum of data points observed. A dynamic environment makes this more challenging than piecing together a puzzle. Imagine that the puzzle picture is constantly changing and so are the pieces. In order to manage and stay ahead of this change, leaders need to recognize the nature of change occurring and its relevance to the response. Is it escalating, de-escalating, or stabilizing? During the COVID-19 pandemic, we experienced periods of escalation, de-escalation, stabilization, then escalation again. We observed crisis leaders reacting to the first escalation with significant countermeasures, but when the second escalation arrived in the fall of 2020, states rarely regressed from phase III back to phase II or modified their countermeasure beyond closing bars early. This proved fruitless in battling the second wave, but may have stabilized economic recovery.

During technical response operations, technicians strive to gather data and enact interventions. It's the leader's responsibility to zoom out, observe the big picture, notice trends, and establish incident objectives and strategy for responding. The disparity between tactical and strategic response may seem contradictory, but it is analogous to points on a curve versus the slope of a tangent on the curve (to use a mathematical analogy). Both are important and necessary. Neglecting or overemphasizing one perspective over the other can lead to deleterious consequences. It is also important to know what you can control and what you can't control during an incident. Too much focus on lagging indicators and trying to solve yesterday's problem detracts valuable resources from leading indicators and tomorrow's solution.

5.2.2 Comprehension

Level 2 is Comprehension of the current situation. This is the synthesis of disparate Level 1 elements (data points). This is taking the colored dots on the maps and understanding their significance with respect to the response goals. This phase is where leaders translate the data into information – sensemaking. To continue with the puzzle analogy, we are no longer looking at individual pieces and placing them in their proper position, but rather observing the picture that is being constructed and comparing the picture with our repertoire of familiar images to comprehend the picture that is being constructed. Keep in mind the picture is still not complete, but we are able to make sense of the limited picture. Assuming we have the most important and critical elements of the picture illustrated, we are able to make informed decisions based on the degree of resolution. When the picture is fuzzy or important elements are incomplete, it is more difficult to discern reality and make good decisions.

There are several elements to comprehension:

- Translating data into information through the compilation of data.
- Calibration of the data based on its reliability, usefulness, and accuracy.
- Comparison of the information with some standard, a library of cases in order to provide context.

Translation of data into information involves the compilation of sometimes disparate data points in order to develop a picture of the situation. These data points are not always from accurate, reliable sources. Data points may consist of individual observations (from a specific perspective), a 911 call, weather data and reports (which are localized and changing), the media, hazardous material detectors which have ranges of accuracy, detection limits and inherent sampling error, electro-optical which show visual conditions, infrared cameras which show degrees of heat variation, or location of resources.

In these examples, there is a standard that must be applied to determine whether the data collected is good, bad, or indifferent to the situation. Leaders should define acceptable versus unacceptable criteria and set action-levels for taking action to change the conditions if they are on path to reach an unacceptable outcome. Table 5.1 lists a few examples:

Table 5.1 Examples of level 2 – Comprehension of data.

Data Input	Acceptable	Unacceptable	Action-level
Observation	Nothing abnormal	A gunman in a school	Potential for criminal activity
911 call	Upset but unharmed person	An intruder or robbery in progress	Description of a crime or imminent harm
Weather data	A rainy day	Flooding	Rain gauges > 5 inches in one hour; stream gauges doubling water level
Media	Reporting on a situation in the mall	Reports of gunfire in the mall	Potential for mass shooting
Hazmat detector	5 parts per million	100 parts per million	50 parts per million
EO camera	White smoke present	Black smoke present	Discolored smoke present
IR camera	No hot spots; black intensity	Hot spot present; white intensity	Indications of heating, grey intensity

These data inputs are just examples. In a real incident, the acceptable, unacceptable, and action levels are specified for the incident. The key is knowing what the acceptable and unacceptable standards are so that good decisions can be made.

In 2014, Freedom Industries, a manufacturer for a coal cleaning substance, had a release of a methylcyclohexanemethanol and glycol mixture from its corroded tanks into the Elk River in West Virginia. This substance made its way into the drinking water system fed by the Elk River and contaminated the water system for over 300,000 residents in multiple counties. The chemical substance was a unique concoction which had little toxicological information available at the time of the incident. No safe drinking-water standards existed for these specific chemicals. Federal agencies rushed to derive data to establish a health standard based on studies conducted by the chemical manufacture. As the response unfolded, the concentration of the chemical was determined to be low, but no acceptable or unacceptable standard was established. In an attempt to assuage residents, the Mayor drank the water and described it as a nuisance but safe. Nonetheless objective data about the exposure effects was unknown. The action level was based on contamination and resulted in flushing tanks and water lines. But how clean was clean? Eventually, the CDC and NIOSH developed an interim standard and the water treatment plan worked with public works to flush the lines and clean the system. Thankfully, the health effects were minor, and the water system was restored. But what if this chemical was more toxic? This case study illustrates several obstacles to making sense of data to support decision-making.

During the COVID-19 pandemic, public-health departments and the CDC rushed to improve their understanding of the novel virus, its effects, transmissibility, mortality, and methods of spread. Data poured in related to cases, hospitalizations, and deaths. But the data was fraught with problems which complicated decision-making. Each state had different standards for determining if COVID-19 was the cause of death or a contributing factor; traditionally, doctors identified one cause of death. Federal standards resulted in paying hospitals more for COVID-19 treatment of patients, so diagnosis became biased to COVID-related in order to improve revenue because hospitals were cancelling elective surgeries and routine visits. Even the case definition took some time to formulate and differentiate from Influenza A and B. Limited test kits, lab analysis capacity, and false positives and negatives affected reliability and accuracy for public-health departments trying to make sense of the data. As the pandemic progressed, testing resources, reporting standards, and case tracking improved. However, even in November 2020, reports from the media of the worst month of the pandemic to date, raised concern of whether we really understood the numbers during the first wave. Back-calculating to estimate the true number of cases based on mortality rate and our understanding of asymptomatic cases, indicated that the first wave was much worse than the data showed, while the second wave which appeared to be worse was not actually as bad as the first.

These case studies illustrate several challenges related to comprehending the data. They also emphasize the importance of managing change and accounting for the dynamic nature of the incident. This is where an understanding of leading and lagging indicators can help decision makers focus on what is relevant to decision-making. Table 5.2 below highlights some leading and lagging indicators from these cases studies.

Notice that the best leading indicators are based on the rate of change. Since incidents are managed over time, we want to influence the rate of change in the direction that we want the incident to move. This is a good segue into Level 3.

5.2.3 Projection

Level 3 is the ability to project future actions of the elements in the environment. It forms the third and highest level of situational awareness. It is achieved through knowledge of the status and dynamics of the elements and comprehension of the situation (Level 1 + Level 2). With this

Table 5.2 Examples of leading and lagging indicators.

Crisis Examples	Lagging	Leading
COVID-19	Positivity rate Cases Deaths	Human temperature heat map Hospitalizations Lag time to get a medical appointment
Freedom Industries Incident	Concentration of chemical in the water	Reports of abnormal odor/taste in water Hospitalizations/recoveries Concentration of total organic compounds at treatment plant
Flammable rail car incident	Observation of a fire Leaking fuel	Pressure relief valves popping Heat tears in the container
West Fertilizer incident	Size of fire Consumption of fire suppressing water	Recognition of insufficient resources to extinguish fire Intensifying fire
Hurricane	Wind speed (category storm) Location	Changing atmospheric pressure (lower) Temperature and humidity

information a decision maker can intervene and change the outcome of an environment. The challenge is identifying and applying a predictive model in order to accurately project into the future. The multitude of influencing elements on an incident result in increased uncertainty.

For example, consider weather forecasts. There is a large body of knowledge, science, data gathering in real time, and modeling, but even the best forecasters fail to accurately and consistently predict the weather, formation of tornados, path of hurricanes, and flooding. Other models that are used in crisis response include:

1) Predictions of the arrival of solar flares (which may affect our satellites and radio communications).
2) Plume diagrams for chemical release.
3) Stream images for flooding.
4) Earth movements around volcanos to alert for destabilizing conditions.
5) Buoys to predict the arrival of tsunamis.
6) Measurements of water levels to indicate droughts.
7) Institute for Health Metrics Evaluation pandemic models.

Many will agree that all models are inaccurate, but they are the best we have in forecasting the future. Before relying on models, decision makers should become aware of the assumptions, limitations, and scope of the model. It is important to build an advisory team of experts who understand the applicability of a model to a set of circumstances.

During the COVID-19 pandemic, many public-health experts relied on the IHME pandemic models to forecast deaths, hospitalizations, ventilator requirements, and cases into the future. These models were based on real-time understanding of mortality and hospitalization rates from Europe and China (which were not reliable or accurate due to under reporting). These models also failed to incorporate the effectiveness of prevention and control measures. Although, models did take these control measures into account in later stages, these measures were not well-calibrated to reality. Rather than rely on these models to make outcome-based decisions, decision makers could use these models to compare alternatives and the associated outcomes.

Similarly, responders can use plume models to estimate the health effects of a chemical on a community but when real-time data is available, these models need to be adjusted for reality to make better decisions. Flood and rain gauges can also be useful in making projections of downstream areas that may flood. But this information needs to be available and timely in order to be actionable.

Modeling products need to be combined with empirical experiences and an understanding of the alternatives with respect to outcomes and risk in order to inform decision-making. This is most apparent when decision makers issue evacuation orders during an approaching hurricane. The lead time required to make a good decision leaves the decision maker with a highly uncertain modeling product. By the time the model gains accuracy, it is often too late to make the evacuation decision and create a successful outcome.

5.2.4 Intervention

The addition of intervention to this model was observed and explained by Burdick at the 2017 Virginia Emergency Management Symposium (VEMS) [4]. Burdick explained that ultimately, a decision will be made, and this decision will affect each element of the model. During the intervention phase, decision makers should look for indications that the decision had the effect that they had anticipated (e.g., change in the trend to de-escalating the crisis). Failure to make a timely decision is also a decision which can have an effect. So, decision makers need to be aware of their window of decision opportunity.

5.3   Relationship between Situational Awareness and Decision-Making

Figure 5.2 illustrates the relationship between situational awareness, internal and external factors, decision-making, and the performance or outcome of the action. When decision makers have a high degree of situational awareness, it has the potential to result in high-quality decisions. It also shows the various factors that affect decision quality and performance.

Situational awareness provides context for making decisions. As situational awareness improves, decision quality may also improve. When a decision maker has poor situational awareness, the decision maker makes uninformed, and often poor-quality decisions. We will elaborate more on critical thinking and decision quality in Chapter 6.

Once a decision is made, it must be implemented. Actions must be performed. Ideally, these actions contribute to controlling the incident, decelerating or reversing adverse conditions, and orienting the incident on the pathway to recovery. Implementation takes time and this should be accounted for in the decision-making process. For example, during an anhydrous ammonia release, the incident commander may decide that the best way to protect the health and safety of people in downwind areas is to evacuate the area. Depending on availability and capacity of transportation routes, this could occur safely and expeditiously. However, in many circumstances, notifying the community of the evacuation order and evacuating the people can take an extended period of time. This timeframe could last longer than the incident and pose additional exposure risks to responders and the people being evacuated. Perhaps a shelter in place order would be more appropriate.

As decisions are implemented, the "state of the environment/system" (in this case the incident) may be affected in some way. Good situational awareness, particularly perception, will detect the changes in the system. As you can see, situational awareness is an ongoing process. Decision makers cannot just conduct one round of "perception," then "comprehend" the situation, and "project" where the situation is heading so that they can make one decision. Instead, situational awareness will continue to evolve as more data is gathered and information is produced. Each decision (or no decision) results in changes to the system. It is dynamic.

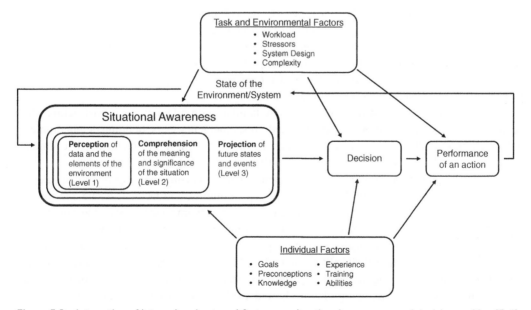

Figure 5.2   Interaction of internal and external factors on situational awareness and decision-making [5, 6].

There are several factors that affect the decision-making process: task and environmental factors and individual factors.

5.3.1 Task and Environmental Factors

Task and environmental factors are external factors such as workload, stressors, system design, and complexity. Workload affects one's ability to conduct situational awareness, make decisions, and perform actions. In a high-workload environment, there is less time available to think through the process, alternatives and project potential outcomes. Workload is determined by the external drivers as well as the available resources to do the work. Previously, we discussed the incident management system and how the incident command system provides a scalable, organizational structure to perform the various tasks associated with incident response. This is one means of controlling the workload by ensuring adequate resources are assigned and equipped to perform tasks. Having facilitated hundreds of exercises, I observed a certain reluctance among incident commanders and leaders to request and assign the necessary resources. There seems to be an inherent bias toward being able to handle a situation without external resources. I have observed this at the local, regional, and state levels of government. The root cause of these cases seems to be (1) lack of relationships and trust and (2) political drivers.

Types of stressors and the effects of stress will be explored in Chapter 8, but it is worth noting that stress results from situations where the demands exceed the resources available. This can occur at the individual or system level. A best practice for managing stress and preventing stressors from adversely impacting the decision-making process is to ensure there are sufficient resources available and these resources are properly organized (i.e., everyone knows their responsibilities). An increase in stress can result in mistakes, premature decision-making, and decision paralysis, among other effects. This creates a "fog of war" situation making it more difficult to successfully respond and manage the crisis toward recovery.

System design includes both the organizational system that enables people to work together, but it could also describe the operating environment. Using the ICS ensures a common framework with unity of command and objectives which can lead to improved decision-making. Whereas disorganization can further cloud the clarity needed for communicating information and making decisions. The external operating environment must be understood. Familiarity with the scene, hazard, incident, infrastructure, people, and potential effects can aid the decision-making process by increasing situational awareness. Pre-incident planning is key to the success of managing a crisis. Knowing the types of hazards and threats, the associated risks, potential impact, and consequences will help decision makers with informed decision-making.

Pre-incident plans help first responders become familiar with a facility, its infrastructure, the location of Hazmat, fire suppression systems, hydrants, emergency exits, building demographics, and much more. When a fire department arrives on scene, these are elements of essential information which can facilitate the execution of the tasks that they need to accomplish.

An evacuation plan for communities near a volcano or wildfire ensures that responders can quickly establish evacuation routes, control traffic, notify the community, and facilitate a safe and timely evacuation. If they have to create this plan during the crisis, much time will be lost and less than adequate plans may be developed in such a short period of time under the stressors of lifesaving.

Complexity of the environment also impacts responders' ability to control and manage the crisis. Complex situations like the BP oil spill in the Gulf of Mexico which resulted from a spill on the floor of the Gulf, with changing weather patterns, currents and tides, distance from land, and potential hazards from volatilizing chemicals and explosive environments, and multiagency coordination made for a complex response operation. These complexities affected decision makers' ability to aggregate timely data, make decisions in coordination, and implement actions that would not result in conflicts such as airspace, watercraft movement, and creation of an adverse impact on other agencies/responders.

5.3.2 Individual Factors

Individual factors include goals, preconceptions, knowledge, experience, training, and abilities. These are specific to the individual involved in the response and decision-making. Unity of command and an incident action plan can help establish and align goals among individuals. Knowledge, skills, and abilities can be predetermined in job action sheets, ICS assignments and validated through a credentialing system. Beyond standardized training and certifications, individuals may have varied levels of training, experience, and preconceptions. These can result in decision bias toward familiar past experiences and may preclude consideration of different situations and outcomes. Decision makers should rely on past experience and training, but should not allow those to limit their options or preconceptions related to the incident.

5.4 Application of the Situational Awareness Model

Improving the skills needed to enhance situational awareness and decision-making takes practice. Exercises are the primary means of practicing these skills and evaluating performance. There are several different types of exercises which can help decision makers apply the situational awareness model and practice decision-making.

Discussion-based exercises include seminars, tabletop, and games. While each of these has a specific purpose, games will provide the best opportunity to practice decision-making, experience the outcomes, and learn from the situation. Discussion-based exercises provide the opportunity for responders to work together within a system, learn from each other, and practice in a forgiving environment.

Operations-based exercises include drills, functions, and full-scale exercises. As the name implies, these are hands-on, performance-based exercises and will usually require more resources but result in more realistic situations where various constraints and stressors can be experienced prior to a real incident.

During the exercise planning meetings, decision makers may provide input to the exercise design and development team regarding the types of decisions that need practice so that the exercise planning team can incorporate opportunities for decision makers to practice with realistic inputs and consequences. Regardless of the type of exercise, leaders will have opportunities to apply this model and practice decision-making. Here are some key focal points for your next exercise:

- How is information flowing to you? From whom? From what systems?
- How valid is the information for supporting decisions? (e.g., raw data, interpreted data, projections, reliability of the source, timeliness, accuracy, etc.)
- How actionable is the information that you are receiving?
- Are decisions being made at the right organizational level (i.e., the lowest level with appropriate authority)?
- What system is used to communicate decisions for implementation and how effective is the system?

- What models are used to project into the future and how reliable are these models?
- What internal and external factors have the greatest effect on situational awareness and decision-making?

Summary of Key Points

- Situational awareness refers to the degree of accuracy by which one's perception of his current environment mirrors reality. It is knowing what is going on around you.
- Dr. Endsley's model of situational awareness has three levels: perception, comprehension, and projection.
- Perception refers to the gathering of data points.
- Comprehension refers to sensemaking, or translating data points into information.
- Projection refers to the process of modeling or estimating a future situation based on trends and existing information.
- The situational awareness model also fits into the decision-making model as situational awareness is used to support a decision which when implemented, affects the environment or system.
- Individual, task, and environmental factors are internal and external factors which impact situational awareness, decision-making, and implementation of the decision.

Keys to Resilience

- **Planning**: Various emergency management plans provide context for how an organization will respond to an incident. Planning begins with hazard identification and risk assessment. This process examines hazard and threats, the risk to the organization or community, and the impact and consequences. This starting point leads to more detailed planning such as an Emergency Operations Plan for general situations, and then specific plans for certain threats/hazards and internal situations. Examples of specific plans include Emergency Response Plans, Hazard Mitigation Plans, Evacuation Plans, Continuity of Operations Plans, Alternate Facility Plans, Pandemic Plans, etc. Solid planning increases situational awareness during the pre-incident phase thereby creating an advanced starting point during response.
- **Training**: Standardized training to specific standards ensures all response personnel can be qualified based on a given standard. However, customized training should also be provided to familiarize personnel with specific local conditions, operating practices, infrastructure, and equipment. Most standardized training achieves limited knowledge-based objectives, sometimes including basic skills application. After analyzing incidents of national significance, I identified that while many of the responders involved in tragic accidents had been trained with the standardized training (knowledge-based), they lacked customized training at higher learning objectives (i.e., application, synthesis, and evaluation). There is a place for both types of training. In combination, responders and decision makers will be better prepared to handle the next crisis. Increasing familiarization of incidents and practicing task execution and decision-making during the pre-incident phase creates muscle memory ensuring readiness.
- **Exercises**: Since the onset of FEMA's HSEEP standard, exercise design, development, and conduct have matured to a new level. This system ensures that exercises are objective-based rather than scenario-based and ensure exercise conduct and evaluation criteria align with the exercise

objectives and the capabilities being evaluated. A well-designed exercise can provide significant value to a community or organization by verifying plans, systems, training, and equipment while identifying areas for improving preparedness. An exercise series that begins with discussion-based exercises and progresses through operations-based exercises enables responders and decision makers to practice in a crawl, walk, run methodology, thereby building confidence, improving situational awareness, and shortening decision cycles.

Application

Consider a recent incident or case study presented in this chapter. Identify examples of how situational awareness was developed based on the following questions:

- What data was perceived?
- Put the data into context with specifics from the situation. Comprehend the situation.
- What was or could be projected with the information provided?
- What decisions were made? What decisions needed to be made?
- What resource did it take to implement decisions? How long did it take?
- What individual, task, and environmental factors affected the decision model?

Search the news for a recent or ongoing incident. Read the story, then synthesize the available information to create situational awareness.

- What data is missing from the story that would help enhance situational awareness?
- Put the data into context. Explain the situation, the severity, the risk, etc. based on the data provided.
- What additional information is needed?
- Estimate what could happen next.
- What decisions need to be made that could drive the incident toward recovery.

Using the case studies identified at the end of a previous chapter, evaluate the decision-making that occurred throughout the incident.

- Map the key decisions on a timeline.
- What were the alternatives at each decision point?
- What decision was made? Evaluate the pros/cons of the decision.
- How would you have decided differently?
- What additional information helped inform your decision?
- If you were facing an incident in circumstances that you had not previously encountered or had little background knowledge, what information, tools, or resources would you want available to help you make a better decision?

References

1 Harrald, J. and Jefferson, T. (2007). Shared situational awareness in emergency management mitigation and response. *40th Hawaii International Conference on Systems Science* (HICSS-40 2007), Waikoloa, Big Island, HI (January 3–6, 2007). Piscataway: The Institute of Electrical and Electronics Engineers, Inc. 23.10.1109/HICSS.2007.481.

2 U.S. Department of Homeland Security (2012). The national strategy for information sharing and safeguarding. https://www.dhs.gov/sites/default/files/publications/15_1026_NSI_National-Strategy-Information-Sharing-Safeguarding.pdf (accessed October 2, 2021).

3 Endsley, M.R. (1988). Design and evaluation for situation awareness enhancement. *Proceedings of the Human Factors and Ergonomics Society Annual Meeting* 32 (2): 97–101. https://doi.org/10.1177%2F154193128803200221.

4 Burdick, B. (2016). Situational awareness: what the science says. Presented at the Virginia Emergency Management Symposium on March 31, 2016 in Newport News, VA. www.vemaweb.org/assets/docs/VEMS16/c3.pdf (accessed June 21, 2017).

5 Endsley, M.R. (1995). Measurement of situation awareness in dynamic systems. *Human Factors* 37 (1): 65–84. doi: 10.1518/001872095779049499.

6 Endsley, M. and Garland, D. (2000). *Situation Awareness, Analysis and Measurement*. Mahwah, NJ: Lawrence Erlbaum Associates.

Further Reading

1 Biferno, M.A. and Dawson, M.E. (1977). The onset of contingency awareness and electrodermal classical conditioning: an analysis of temporal relationships during acquisition and extinction. *Psychophysiology* 14 (2): 164–171. doi: 10.1111/j.1469-8986.1977.tb03370.x.

2 Klein, G., Moon, B., and Hoffman, R.R. (2006). Making sense of sensemaking 1: alternative perspectives. *IEEE Intelligent Systems* 21 (4): 70–73. doi: 10.1109/mis.2006.75.

6

Decision Theory

In Chapter 5, we presented a model for situational awareness and how observations, data, and other inputs can be perceived and comprehended so that a future projection can be made about the situation. This model is influenced by individual and environmental factors, and often unfolds in a dynamic uncontrolled environment. Nonetheless, decision makers must utilize situational awareness to improve problem definition in order to solve the right problem with sound decision-making. The next phase of the decision-making process involves critical thinking.

Learning Objectives

After reading this chapter, you will be able to:
- Apply Endsley's model for situational awareness.
- Apply critical thinking skills in solving decision-based problems with one or more of the models presented in this chapter.
- Understand how time constraints affect decision quality.
- Synthesize new information to refine situational awareness and improve decision-making.
- Identify the role of expert advisors in formulating courses of action which support decision-making.
- Apply decision theory in several group exercises to develop courses of action, establish decision criteria, and make decisions.

6.1 Critical Thinking

Critical thinking is an essential skill needed to improve situational awareness. It requires clear, unbiased thinking and rationality to form a sound judgment. While there are various theories on the principles involved in critical thinking, we will present one example below that leads into the decision-making process and apply these in the situational awareness model to illustrate how critical thinking principles are necessary. Critical thinking is defined by the National Council for Excellence in Critical Thinking (1987) in a statement by Michael Scriven and Richard Paul, presented at the 8th Annual International Conference on Critical Thinking and Education Reform (1987) [1]:

"**Critical thinking** is the intellectually disciplined process of actively and skillfully conceptualizing, applying, analyzing, synthesizing, and/or evaluating information gathered from, or generated by, observation, experience, reflection, reasoning, or communication, as a guide to belief and action."

Situational awareness provides the information generated from observations, detection, sensing, perceiving, and communication. This information must be applied to define the problem or set of problems that the decision maker is trying to resolve. The set of data and information must be analyzed to determine the credibility, reliability, accuracy, and precision. It must be synthesized from multiple sources: standards, public observations and perceptions, responders, government authorities and experts, private sector stakeholders, equipment, information systems and technology, and models. Those charged with decision-making must evaluate each element of information, synthesized information and trends, and courses of action in order to form a judgment and make a decision.

The critical thinking process must be unbiased and open to new information, opposing alternatives, reconstructive analysis, assumption clarity and verification through testing, and sincere desire to reach ground truth. In a time-constrained, dynamic environment, it is impossible to frame 100% of the information in a perfect, rational, and logical mosaic, so decision makers must accept the imperfections and ambiguities while vectoring the process toward optimally accurate situational awareness to best inform the decision-making process. Ultimately, the process of reaching logical conclusions, solving problems, analyzing factual information, and taking appropriate actions based on the conclusions is called decision-making.

Let's reexamine our situational awareness model and apply critical thinking principles to each level.

6.1.1 Perception Applied

Under the perception level, we manage data inputs. We recognize and monitor data inputs. These can be observations, sensor outputs, reports, photos, or metrics. These inputs are typically organized and displayed in some manner such as a common operating picture, planning meeting reports, radio transmissions, or signal output displays. With so many signals and data inputs enabled through Bluetooth, WiFi, Supervisory Control, and Data Acquisition (SCADA), Bodycams, satellite photos, Unmanned Aerial Vehicle (UAV) cameras, Geographic Information System (GIS) displays, word clouds, infographics, and more, the volume of data inputs can become overwhelming for decision makers. Additionally, while each of these elements may have some importance to someone, all elements are not necessarily important or necessary to create situational awareness for the decision maker. The challenge is distilling useful data into a visual format that enables decision-making. Too much data can overload the decision maker; while not enough data may stall the decision maker due to a feeling of insufficient data. The cognitive limits of the human brain are five to nine simple elements. Information overload can impair decision-making.

Like the incident command system, where the ideal number of people reporting to a decision maker is seven (plus or minus two), situational awareness tools should aim to provide a similar volume of data displays to ensure the system is within the human cognitive limits. Failing to deal with information overload results in multitasking. Multitasking impairs productivity. Each time the brain moves from one piece of information to another, the brain must stop the thinking process, change to the next, start back up and reorient, then process the information. Studies have shown that a person can improve productivity by performing two simultaneous tasks, but when adding more tasks, productivity decreases due to the time lost to switching information elements and reorienting the brain to the next element. While technology may be capable of providing increasing information, we lack the ability to process this information efficiently. This leads to delayed decision-making and lower accuracy. Nonetheless, current technology helps us achieve perception.

6.1.2 Comprehension Applied

The second level of situational awareness according to our model is Comprehension. Data alone is not particularly useful or meaningful. In the context of a crisis, it is important to accurately define the problem(s) in order to put the data into the proper context and engage in analysis. This is where critical thinking becomes more important. Those involved in the decision-making process must engage in putting information into context, synthesizing information from multiple sources, analyzing the validity of the information, and evaluating the information to form a judgment. Let's walk through a real-world example.

Example: A hazardous material response team arrives on scene at a train derailment. The first step is to perceive data inputs so that they can be put into context and define the problem. Defining and solving the wrong problem could have deleterious consequences.

Defining the problem: Critical data includes determining the cargo on the train and whether it is hazardous. If not, the responders can quickly transition into recovery, clean up, and avert any public safety concerns. However, if the cargo contains hazardous materials, there could be health and safety concerns for responders and the public, as well as potential environmental concerns. Consider the follow scenarios:

- **Scenario A:** Tanker cars containing fuel were involved in the derailment but the cars are not leaking, nor on fire. The situation is controlled and contained. Responders can implement safety measures and transition to recovery. The problem is simply a derailed car that needs to be put back on the track or emptied of its contents and tested before reuse.
- **Scenario B:** Tanker cars containing fuel were involved in the derailment and are leaking but not on fire. This presents a hazard to the environment, potential inhalation hazard to responders and the public, and possible explosive atmosphere concerns which result in restricting any spark-producing equipment from the immediate area. The problem is a leaking tanker car that needs to be plugged to stop the leak and recover the car, as well as an uncontained hazard in the environment which must be cleaned up.
- **Scenario C:** Tanker cars containing fuel were involved in the derailment and are leaking and are on fire. This presents several hazards to the environment, responders, and the public. The problem is a leaking tanker car, which is on fire with a combustible hazard. This situation presents an imminent hazard to responders and those in the immediate area, especially if the situation is escalating. The solution is significantly different from the other situations; in this case, the solution may be to disconnect nearby rail cars to create significant separation distance and evacuate the area.

The situations presented above were based on visual observation such as the United Nations number (a four-digit number that identifies hazardous materials and articles in the framework of international transport) on the side of the tanker car and presence of leaked material and fire. But additional sources of information may become available that support the problem definition. This could include measurements of hazardous materials at the incident and downwind in nearby residential areas, infrared video from a UAV showing liquid discharge or heat pockets on the car that may indicate potential for fire, or weather reports identifying a change in wind direction with a passing front. Each of these elements need to be synthesized to provide a full context and then analyzed for reliability and accuracy. Consider the following situations:

Situation A: The hazardous material response team measured 15% of the Lower Explosive Limit (LEL) on their Hazmat gas sensor near one of the tanker cars. This measurement needs to be analyzed in context of what is normal, abnormal but acceptable, and abnormal but unacceptable. Other critical elements of information which support reliability and accuracy include the following: was the detector calibrated prior to team use and found to be within acceptable tolerances? Is the team applying the appropriate correction factor for the chemical or mixture involved in the incident since the detector is calibrated with a different gas than the hazard present on scene? Is the team trained and proficient with the detector? While the decision maker may not have all of these details, they need to ensure that there is a reliable system in place to uphold the integrity of measurements made and report which influence their decision. In this case, 15% of the LEL exceeds the action level of 10%. This is abnormal and unacceptable. There is a potential for an explosive environment. The decision maker needs to ultimately understand the significance of the excessive measurement. If this were simply a hazard substance that presented an inhalation hazard in the immediate area of the incident, then the decision to wear a respirator might suffice. Whereas the explosive atmosphere may be an indication to evacuate the area.

Situation B: The strike team deploys a UAV equipped with an infrared camera. Infrared cameras measure heat gradients. If calibrated properly, the color or intensity of the gradient can be used to determine the temperature of the object. If not, the gradient may still be useful in indicating a problem. By nature of the outside air, inside vapor, and inside liquid, temperature gradients will exist. A leaking tanker car could be easily identified using this method. The lack of accurate temperature calibration would not matter in the problem definition, since the uncontained fuel in the environment is a problem. However, if a pressure relief valve blows and the infrared camera detects an intensifying heat gradient, this might indicate the potential for the container to catch fire or explode under vapor pressure. Again, an accurate measurement of the temperature may not be needed to make an accurate observation with this device.

Situation C: The National Weather Service forecast indicates that a cold front will pass through the area. Current wind measurements show that the winds are out of the south at 5 knots. However, the forecast for the next hour is that the temperature will drop and the winds will shift to the east at 15 knots. This information may not affect the response operation. But there are several considerations related to critical thinking. The NWS is a credible source of information. The accuracy of their forecast can be determined by reading the forecasters description and considering historical accuracy for similar situations. While decision makers may not know the exact time of the wind change, they can be certain within one to two hours based on the forecast. Other contextual information might include the location of the responder staging area and how the wind shift could impact them or downwind residents or businesses. With some lead time, responders could safety evacuate an area or issue a shelter in place order. Additional modeling might show that the higher wind speed will dilute any volatilizing chemicals such that the downwind hazard distance is minimal.

It should be clear how different types of information are necessary to improve situational awareness and how small changes in one element could render other elements less important to decision-making. Other considerations in our highly technical context include understanding the capabilities and limitations of the equipment that we use to provide information. In our example, we highlighted a hazardous material detector. These detectors are not capable of detecting all hazards; so, a team's detector could show no hazard detected but a hazard (which the detector is incapable of detecting) could be present. This is commonly referred to as a "false negative" result. This equipment also has limitations on how low of a concentration it can detect as well as interferences which can create a "false positive" result.

During the COVD-19 crisis, situational awareness was very challenging due to similar issues. Antigen testing was highly inaccurate. Doctors could rely on a positive result, but not a negative result. Even PCR testing was imperfect. Case definitions were broad at the beginning of the crisis and most of the reliable data included deaths and hospitalization. Cases were mostly based on clinical diagnosis which overlapped with influenza symptoms and could not be confirmed with PCR testing during the early stage of the crisis due to lack of testing resources. Antibody studies showed that the true number of cases was likely 10 times the number of reported cases based on PCR testing alone. Crisis decision-making was further complicated by the confluence of multiple public health crises: COVID-19, 400% increase in suicide rates and increases in depression which resulted from isolation, unemployment, stress, and economic hardship, and drug and alcohol abuse. The right course of action for one crisis may also impact other growing crises.

Situational awareness tools may provide only data (i.e., level 1 perception) but when the data is put into context, it is translated into information. Once information is available, a person or system must determine whether it is useful information for decision-making. Dashboards, heat maps, trending software, and other tools can help make sense of data to achieve comprehension. Finally, consider the source of the data when determining its reliability, authority, and validity for informational purposes. Identify underlying assumptions, limitations, tolerances, and information gaps. While a complete and comprehensive picture of the situation is desired, decision makers will only have the information that is available at the time.

6.1.3 Projection Applied

Projecting or predicting the future is challenging but necessary to provide the full context and understand the speed and necessity of decision-making. Current situational awareness tools are not especially adept at predicting the future in real time. Several examples of predictive models include weather forecasts (NOAA), plume models, evacuation modeling, and fate and transport models of contaminants in water (USACE Engineering Research and Development Center).

Most of the existing models require user input to project into the future and estimate the impact. This could serve as a distraction from timely decision-making; therefore, it may be helpful to conduct pre-planning and analysis for specific incidents and incorporate the projections into pre-incident or hazard specific plans for future reference.

Ultimately, someone with skill and understanding of the capabilities and limitations of models will need to interpret or analyze the output and evaluate its usefulness. Consider the multiple spaghetti models used in hurricane track and landfall forecasting. These are rarely accurate but still useful in issuing evacuation warnings, understanding the timeframe for decision making, and projecting the impacted population.

One of the most useful applications for projections and models is in comparing alternatives and projecting resource allocation needs. During the COVID-19 pandemic, decision makers at the state and federal level utilized various models to estimate the number of cases, hospitalizations, and deaths in the near and long term. Near-term projections assisted in allocating and establishing surge capacity resources such as ventilators, establishing alternate care centers, and PPE distribution. Long-term projections were used to compare various mitigation measures and associated outcomes. While not a perfect science, these projections helped contextualize the severity of different alternatives and drove mitigation measure decision-making. A major limitation of these models was that mitigation effects were estimated on a macro-scale, and these models did not incorporate secondary or tertiary effects of the mitigation measures associated with lockdowns such as depression, anxiety, suicide, or economic impact. This recent example illustrates the complexity of critical thinking synthesis, analysis, and evaluation of alternatives.

6.1.4 Problem-Solving

Critical thinking applied to situational awareness leads to decision-making. As information is analyzed, synthesized, and evaluated, problems are defined with more clarity and solutions are formulated to resolve the problems. In a simple situation, a problem can be defined with a high degree of clarity and the solution may be obvious. In a complex situation, there may be more than one problem with multiple solutions, secondary effects, or problems that will result from some solutions, or the problem-solving process may become iterative. The key to problem solving is accurately defining the problem or problem set along with the influencing issues. It may be helpful to analyze the problem from different perspectives to ensure that the multiple dimensions of the problem are considered. There are multiple methods to problem solving such as:

● PDCA: Plan-Do-Check-Act [2]
● OODA loop: Observe, Orient, Decide, Act [3, 4]
● Lean eight-step process [5]

Each of these processes leads to decision-making. Let's take a closer look at the eight-step problem-solving process as it has several unique aspects which can result in improved decision-making. During a crisis, the key to applying a problem-solving model is practice. It should become a second nature.

1) **Define the Problem:** During a crisis, the situational awareness model will provide multiple problems which require intervention. The Comprehension and Projection phases of the model will be necessary to formulate the problem definition.
2) **Clarify the Problem**: Critical thinking involves synthesis of disparate data and information, analysis among stakeholders to include diverse perspectives, identification of gaps in information, and evaluation of the information can lead to a judgment about the problem.
3) **Define the Goal:** It is easy to identify problems but this step begins the process of formulating solutions by defining a goal or end state. During a crisis, this may entail recovering to a normal, stable state. The goal or objective may also include a schedule or desired result.
4) **Identify Root Cause of the Problem:** Solving a problem may lead to a temporary fix, but unless you understand why it is a problem, it may continue to manifest in similar or other

manners. Asking "why is it a problem?" and "how did we get here?" begins to reveal underlying causes which must also be addressed as part of the problem to prevent recurrence.

5) **Develop an Action Plan:** This step begins with brainstorming ideas in order to develop solutions. Each idea must undergo testing and evaluation to ensure the contemplated action will be effective in resolving the problem. The end result of this step is several courses of action for decision-making.

6) **Execute the Action Plan:** Once a decision is made, the plan must be implemented. Leaders should always verify implementation with "Did you do it?" and "Was it done?"

7) **Evaluate the Results:** A plan perfectly executed may be effective, but if the problem definition was off or the plan was not perfectly aligned with the problem and root cause, the implementation of the plan may not have been effective. Therefore, it is important to always have a measure of performance and measure of effectiveness to assess actions taken. Ultimately, the crisis leader needs to resolve the problem.

8) **Continuously Improve:** Ongoing operations provide an opportunity to capture lessons learned and make adjustments in how actions are implemented. Good documentation of decisions, actions, and effects should be included in an after action report to help improve performance in the future.

These problem-solving processes overlap with situational awareness and decision-making. We are just viewing different aspects of the process in this section. Before moving on to the decision-making process, let's expand our discussion on developing alternatives or courses of action.

6.1.5 Courses of Action

As a crisis unfolds, one significant resource for decision makers is a team of experts that can make recommendations to the decision maker. If a decision maker has good alternatives to choose from, they are more likely to make a great decision. A good leader asks their team "what do you recommend?" to fully utilize the brain power and capacity of those reporting to the leader. If the ultimate decision maker attempts to comprehend, project, and formulate alternatives, they will quickly become overwhelmed. Instead, a crisis leader must rely on their resources to analyze information, develop recommendations, and communicate them effectively so that the best decision can be made.

There are two common approaches to this process: parallel and serial. The parallel approach illustrated in Figure 6.1, shows how an analyst seeks out information from a broad range of sources, conducts analysis, and presents alternatives to the decision maker. In this approach, the analyst considers all available information, including areas outside of their expertise. This would be analogous to the public health official analyzing economic and security impacts of their recommendation to the decision maker, as well as the senior law enforcement official analyzing economic and health impacts associated with their recommendations. This could result in some biases or multiple recommendations. An alternative approach is to follow the serial process where the experts in each area research and analyze only the information within their domain of expertise, develop recommendations, and present the best approach to the decision maker. In the serial process, the decision maker would receive a recommendation from each senior advisor and perhaps choose one of the recommendations or some combination of those recommended. The strength of the serial approach ensures that the decision maker is getting the best course of action for each expert, typically results in fewer alternatives and improves the likelihood of a great decision.

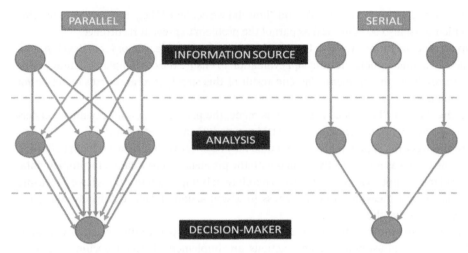

Figure 6.1 Comparison between parallel and serial information processing [6].

Human cognitive limits require relevant and actionable information with fewer inputs and choices. More options and inputs take longer to compile, analyze, and decide. Therefore, it is important that decision makers surround themselves with competent advisors that are capable of gathering relevant information, analyzing the information in a timely manner, and communicating the alternatives effectively to facilitate decision-making. When advisors are not competent, they tend to spend excessive time researching and considering alternatives beyond their scope of expertise. Expert advisors who are focused on their discipline are best able to support decision makers under time constrained conditions. In this next section on decision-making, we will consider how time and quality of information affects decision quality.

6.2 Decision-making

After formulating courses of action, it is time to make a decision. There are three general cases we should consider in decision logic:

- No alternatives
- One alternative
- Multiple alternatives

The first general case is the status quo, where no alternatives or courses of action are established. We need to recognize that "no decision" is a decision. It's a decision to not formulate an alternative, and therefore accept the status quo. A good leader will recognize that they are deciding to choose the status quo over some alternative that could be developed. Crises do not manage themselves; leaders manage the crisis whether they realize it or not. If a leader is not satisfied with the current status or projected outcomes, then it is incumbent on the leader to formulate an alternative or direct their team to do so.

The second general case is when only one alternative is formulated which is different from the status quo. In simple situations, this may improve the quality of the situation and drive the crisis toward recovery. But in complex situations, one alternative may be too simplistic for the circumstances. It may not address the totality of the situation. Nonetheless, the leader has two alternatives: the status quo or the newly proposed alternative which may likely improve conditions but not completely.

The third general case is when multiple alternatives are developed and present to the leader. This case has the highest likelihood of improving decision quality, especially when multiple experts were involved in independently developing the alternatives. The leader is more likely to have several good alternatives to choose from; therefore, the quality of the decision and outcome is likely to be a high-quality decision. Establishing these conditions begins during the pre-incident phase by ensuring adoption of an incident management system that assigns roles and responsibilities to experts (to include their alternates), training personnel, and practicing during exercises. Time and resource constraints affect the team's ability to create these conditions for success.

6.2.1 Resource Constraints

Rural communities experience the greatest resource constraints: people, talent, financial, and equipment. Smaller communities typically have smaller government organizations and volunteers, or part-time responders available to support the community. Availability of trained, educated, and knowledgeable personnel is a limiting factor compared with large, urban areas with a larger pool of personnel to access. Reachback or virtual support can help bridge the knowledge gap. Financial resource limitations also affect access to the latest technology, equipment, and infrastructure. While every community experiences some extent of resource constraints, response capacity and limitations should be understood through a gap analysis process during the pre-incident phase. Communities may be able to leverage external resources where significant gaps exist through access to state or regional teams, private sector or contract resources, lease, or rental agreements, and grants. Mutual aid agreements and Memoranda of Understanding are commonly developed between parties to access and leverage resources.

6.2.2 Time Constraints

Time constraints greatly affect decision quality. As we addressed in Chapter 5, situational awareness is a process that takes time and resources to develop and optimize. Initially, crisis leaders will lack full situational awareness, but it will improve based on access to knowledge, observations, and sensors involved in the situation. The time it takes to increase knowledge of the situation, comprehend the situation, and project into the future based on the dynamics affects the decision cycle. Sensing or searching requires personnel and equipment that can perceive data about the situation. Knowledgeable personnel need to interpret and comprehend the data to make sense of it. Critical thinking aids in validating, confirming, and analyzing the information to improve clarity. Alternative development and decision quality can be affected by the quality and clarity of this information.

Imagine a low-resolution picture informing a sniper on potential enemy targets during a battle. Without adequate clarity, there is a risk of missing the target. But as the resolution of the picture increases, identifying marks can be discerned, obstacles can be identified, and mistakes can be averted. In this example, we are informed that the resolution is increasing. But in a crisis, more information does not necessarily improve the clarity of the situation; it could be more noise. Critical thinking is necessary to validate the information for reliability and accuracy. Therefore, understanding the component of time is an essential element of decision-making.

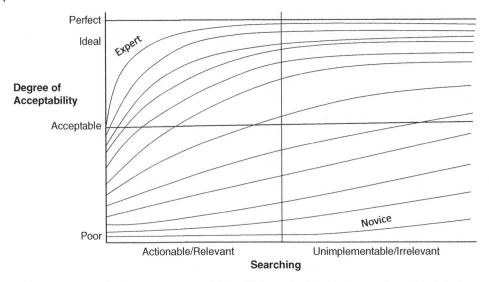

Figure 6.2 Relationship between acceptability of information based on expertise and search time.

Clarity can be derived from knowledgeable personnel and systems that integrate data in order to provide useful information. Without knowledgeable personnel or systems integration, time spent searching for the right information increases. Leaders supported by knowledgeable personnel (i.e., experts) are more likely to make higher-quality decisions because they spend less time searching for the right information. Expertise influences the degree of acceptable information provided in support of decision-making. Figure 6.2 illustrates the relationship between decision quality, time spent searching for more information, and the informant's level of knowledge and expertise.

6.2.2.1 Decision-making: No Time Constraints

Decision makers are rightly reserved about making decisions when they lack sufficient information to make a high-quality decision. As the downside potential associated with a "wrong decision" increases, decision makers become more risk averse. No one wants to make a life-and-death decision when they lack clarity and good situational awareness; the risk and liability is high. So, decision makers will rightly seek more information to improve their situational awareness in hopes of improving the decision quality.

Consider for example, launching a space vehicle. Space vehicles are highly complex. There are numerous factors that can affect the successful launch (e.g., reliability of the systems, weather, controls, demonstrated repeatability, etc.). Having the right information and being certain of the launch vehicle status and environmental conditions are critical to a successful launch. Under routine situations, delaying a launch may face political or some economic pressures. But what if the launch is necessary to resupply the International Space Station with lifesaving supplies? Waiting for normally acceptable conditions could result in loss of life. Decision makers would need to assess their tolerances for acceptable launch conditions and compare these risks with the risk of not launching the mission.

Figure 6.3 illustrates how quality of decision is impacted by quantity of relevant information (which may be related to time). When time is not a factor, gathering additional information improves the quality of the decision but at some point there is a diminishing return. With little to

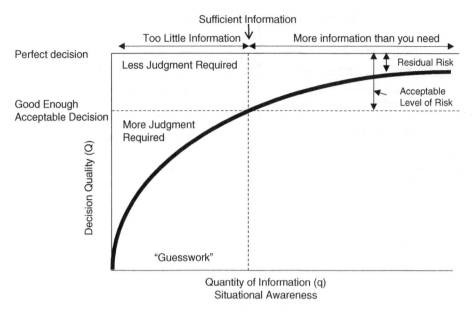

Figure 6.3 Decision-making without time constraints [6].

no situational awareness (or information), decision makers may be guessing when making a decision; more information and judgment is required to attain an acceptable (good enough) decision. As situational awareness increases, a point is reached where sufficient information is attained to make an acceptable decision. When situational awareness increases beyond this point, the residual risk continues to decrease and decision quality should continue to improve, approaching but never reaching the perfect decision.

This figure makes the following assumptions:

- Decision Quality (Q) improves with Information Quantity (q).
- Slope is always positive: $dQ/dq > 0$.
- Gains of Q are great when q is small; gains of Q are less as q increases.
- There exists some "Perfect Decision" that is approached as q increases but is never quite reached.
- More information always yields better decisions.

6.2.2.2 Decision-making: Time Constrained

Under time constraints, failure to act in a timely manner (paralysis by analysis) will result in worsening quality decisions as more information is gathered. Decision makers must combat the temptation to continue to gather information in an attempt to avert risk (right side of the graph). Decision makers must define an acceptable level of risk so that they can make timely, imperfect, yet acceptable decisions. Figure 6.4 illustrates this dynamic. Unlike Figure 6.3, Figure 6.4 shows how decision quality decreases with too much information; decision makers that exhibit risk aversion may find themselves operating in this region (on the right side of the peak). While they may still make an acceptable decision, they are losing time to implement the decision once it is made. Good leaders should aim for the center of the red-dashed matrix. This is the region associated with sufficient to optimal information where decision quality continues to improve beyond "acceptable" toward the best possible decision.

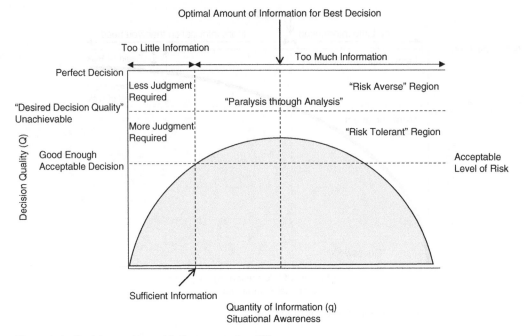

Figure 6.4 Decision-making with time constraints [6].

Figure 6.4 embraces the following assumptions:

- Decision Quality (Q) improves with Information Quantity (q) when q is sparse.
- Q decreases with increasing q when q is abundant.
- Slope is not always positive (i.e., dQ/dq can be (+) or (−)).
- There exists some Maximum Quality of Decision that cannot be improved upon by either more or less Information at the peak of the curve (i.e., dQ/dq = 0).
- There exists some unachievable "Perfect Decision."

6.2.2.3 Decision-making: Making Good Decisions

The formula for making good decisions begins with creating optimal conditions prior to an incident. Decision makers need relevant, reliable, and timely information to make good decisions. Keys to success include the following:

- Team of technical experts, knowledgeable and trained in the hazards/threats causing the crisis.
- Real-time information systems that aggregate, display, and analyze relevant data points; this may include common operating pictures and the use of Community Lifelines.
- Analysts that can make sense of the data, apply critical thinking for validation, and possess the ability or resources to make realistic projections.
- Organizational structure such as ICS with a system for tracking, reporting, collaborating, and developing courses of action.
- Multiple good alternatives from which to make a decision.
- Pre-incident planning and documented procedures.
- Practice through exercises.
- Effective and streamlined communication systems needed to notify responders of a decision so that they can implement it in a timely manner.

6.2.3 Decision-Making Models

While there are multiple decision-making theories and models, this chapter will present two models that are very different but equally valid approaches which may be helpful to leaders who are looking for methods for evaluating alternatives and making a decision: an objective model and a subjective model.

6.2.3.1 Data Quality Objective Model

One framework for establishing decision criteria, boundaries, and objectives is the EPA's data quality objective process. The DQO process was designed for application in cleaning up Superfund sites and determining whether the site was remediated to an acceptable level. While this process would normally unfold over a long period of time, there is some applicability of the process to decision-making. The DQO process is a series of logical steps to establish criteria for collecting data of sufficient quality and quantity to support objective decision-making. While the EPA model relies on quantitative data, the framework of this model is helpful in understanding the decision-making process and the inputs, criteria, and conditions required to make objective decisions. During crises, there may not be sufficient time to employ this model to the fullest, but the framework is helpful in establishing objectives, boundaries, and decision criteria for those that supply decision-makers with information.

---

1. State the Problem
Summarize the contamination problem that will require new environmental data, and identify the resources available to resolve the problem; develop conceptual site model.

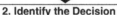

2. Identify the Decision
Identify the decision that requires new environmental data to address the contamination problem.

3. Identify Inputs to the Decision
Identify the information needed to support the decision and specify which inputs require new environmental measurements.

4. Define the Study Boundaries
Specify the spatial and temporal aspects of the environmental media that the data must represent to support the decision.

5. Develop a Decision Rule
Develop a logical "if…then…" statement that defines the conditions that would cause the decision maker to choose among alternative actions.

6. Specify Limits on Decision Errors
Specify the decision maker's acceptable limits on decision errors, which are used to establish performance goals for limiting uncertainty in the data.

7. Optimize the Design for Obtaining Data
Identify the most resource-effective sampling and analysis design for generating data that are expected to satisfy the DQOs.

Figure 6.5 Data quality objective decision-making process [7]. *Source:* United States Environmental Protection Agency. (2000, January). Data Quality Objectives Process for Hazardous Waste Site Investigations. Washington, DC 20460.

The following steps, summarized in Figure 6.5, describe the process:

Step 1 requires the decision maker to define and state the problem. For example, there may be a continuous release of Anhydrous Ammonia from a pipeline in a community. People exposed to a high dose may become incapacitated.

Step 2 is to identify the goal or objective related to the data that will be collected (both qualitative and quantitative) in order to make a decision. For example, the objective may be to protect the public, save lives, and prevent exposures that would have an adverse health or safety impact (i.e., hospital visits).

Step 3 is to identify the information inputs. In this case, there may be several inputs such as pre-incident plans, plume modeling, detection results, reports of people experiencing exposure, weather conditions, and environmental monitoring. Some qualitative information might include the population and demographics of people affected by the release within the hazard area. Other inputs could include the topography, businesses, schools, hospitals, subdivisions, and agricultural areas that could be affected.

Step 4 involves defining the boundaries of the problem. These include spatial and temporal limits such as an area of land and timeframe over which data will be collected and the decision will be made.

Step 5 develops the analytical approach to include parameters of interest. This will be based on the capabilities and capacity of responders to collect and integrate information. In the most basic and limited case, an incident commander might utilize an estimated volume of gas released to estimate the area with highest impact using a calculation, model, or input from a regional Hazmat team. A more sophisticated approach might involve utilizing reports from the public on the severity of their exposures or downwind detection results. If this information is available, then the decision maker might compare detection results with Emergency Response Protection Guidelines or some other action level to determine severity of the hazard. A Hazmat response team might utilize a plume model to determine their sampling strategy.

Step 6 specifies the performance or acceptance criteria. This is where the decision maker needs to accept risk in determining how much data and information is sufficient to rely upon in making the decision and consider the risk of not making the decision. This is also based on resource constraints. The leader will have to decide how much of their resources will be allocated to collecting information to inform their decision-making (e.g., wide area detection surveys) versus implementing the decision (e.g., evacuating an area). The hazard condition is changing with time, wind speed, etc. but at some point, a decision needs to be made to evacuate, shelter in place, or do nothing. Evacuation could put people at risk of direct exposure. Sheltering in place is a temporary measure that can be more effective if residents seal windows and doors and turn off their HVAC, but over long time periods this could result in an accumulation of the hazard within homes (which are permeable) where people are trapped. Doing nothing might result in some exposures depending on the environmental conditions but could enable more resources to be invested in stopping the leak. If the objective is to prevent incapacitation (i.e., hospitalizations), the leader may accept minor exposures while focusing resources on preventing and controlling severe exposures.

Step 7 in this model is to employ resources in implementing a plan to collect the information necessary to make the best decision. In our example, the optimal outcome minimizes or prevents high-dose exposures which could result in hospitalizations. This is an iterative process and may involve parallel decisions. Starting with the problem and decision objective sets the conditions for how leaders allocate resources to collect the right information needed to support the decision. Establishing decision criteria based on acceptability is equally important as there may be cases where information is provided but acceptance criteria cannot be established. Consider the Freedom Industries case study from previous chapters. The chemical involved in the incident did not have an acceptable standard for safety purposes so the drinking water sampling did little for health and safety decision-making, but real-time correlation between the severity of exposure and concentration in the water could have improved decision quality.

6.2.3.2 Subjective, Empirical Model

Professional judgment will ultimately play a role in decision-making, especially when little information is available to the decision maker. The decision maker will rely on their experience with similar situations, their knowledge, and input from others. When navigating a complex, chaotic, and uncertain crisis, it may be difficult to obtain the information needed to create certainty. The COVID-19 crisis is a good example of a global crisis where situational awareness was lacking for months given the long incubation period, wide range of medical effects, lack of adequate testing and timely analysis, and knowledge related to the efficacy of the prevention and control measures. So how can a leader navigate this situation and make good quality decisions?

Jim Collins' book *Great by Choice* provides solid guidance based on empirical evidence and analysis on thousands of companies that thrived throughout decades of uncertain, chaotic, and complex periods [8]. His team of researchers identified companies in different industries that significantly outperformed their competitors over a 30-year period despite economic recessions, regulatory changes, technology advancements, labor issues, and other factors. His team identified several factors that contributed to the success of the "10Xers" (they outperformed their competition by 10 times in growth):

- Fanatic discipline
- Empirical creativity
- Productive paranoia
- Specific, methodical, and consistent recipe.

The first attribute is "Fanatic discipline." Leaders of the 10Xers did not make erratic changes to their business objectives, growth, or mission. They exercised fanatic discipline and steadiness in their leadership approach. Collins reminds us that "True discipline requires the independence of mind to reject pressures to conform in ways incompatible with values, performance standards, and long-term aspirations ... They're capable of immense perseverance, unyielding in their standards yet disciplined enough not to overreach." "They started with values, purpose, long-term goals, and severe performance standards; and they had the fanatic discipline to adhere to them."

Second, he attributed empirical creativity to the 10Xers. Empirical creativity is the ability to balance empirical evidence with creative approaches. This includes adoption of best practices, lessons learned from other incidents, and use of innovative technologies. Crisis leaders should be familiar with case studies, best practices, and lessons from after-action reports. This equips them to use empirical evidence in decision-making. Empirical creativity is a departure from conventional wisdom to set a different course during times of uncertainty. These leaders did not primarily look to what other leaders were doing, or to what pundits and experts say they should do. Instead they looked primarily to empirical evidence. During the COVID-19 crisis, early adopters applied traditional infectious disease prevention and control measures such as physical distancing, handwashing, isolation, quarantining, and wearing masks. In early December 2020, the CDC revised its quarantine guidance based on empirical evidence in order to improve compliance and manageability.

Third, productive paranoia is the regular state of conducting "what if" analyses, testing decisions, and alternatives to estimate the outcome. This also includes an understanding of risk. Determining the acceptable risk, or "death line," referring to the unacceptable risk associated with climbing Mt. Everest at an altitude with insufficient oxygen supplies, adverse weather forecasts, and recognizing unacceptable conditions in time to act. This also requires planning for the unthinkable, black swan event. This understanding of improbable but high-consequence events helps leaders keep perspective during a crisis. During the COVID-19 crisis, lack of hospital capacity and ventilators was the unacceptable risk. The federal government turned hotels and stadiums into hospital rooms. They activated the Defense Production Act to manufacture ventilators, and closed international air traffic.

Zooming in and out ensures that leaders are not caught in the details which enables them to continue monitoring the big picture. Collins explains, "10Xers constantly considered the possibility that events could turn against them at any moment. Indeed, they believed that conditions will – absolutely, with 100 percent certainty – turn against them without warning, at some unpredictable point in time, at some highly inconvenient moment. And they'd better be prepared." During the COVID-19 crisis, effective leaders were able to zoom out from the public-health crisis of the disease and forecast the looming mental-health crisis associated with isolation, as well as the economic impact associated with lockdowns. This revealed the complexity of the crisis and showed how effective actions in one domain opposed effective actions in another domain, resulting in compromises based on relative risk.

When following the ICS, these "what if" analyses should be assigned to the Planning section to evaluate and manage risks. One particularly relevant finding to crisis leaders in Collins' analysis was the recognition that effective leaders were able to detect changes in the risk profile in a timely manner. They had the presence of mind to ask the question, "how much time before the risk profile changes?"

Finally, successful leaders applied a SMAC recipe within their organization: Specific, Methodical, and Consistent, and found, "the signature of mediocrity is not an unwillingness to change; the signature of mediocrity is chronic inconsistency." There is a delicate balance between change, continuous improvement, and being inconsistent. Written procedures and training are essential to knowing the job and implementing the response. Reinventing a response team's procedures, or implementing a new organizational system that has never been practice, puts the response team at a disadvantage.

It should be no surprise that companies that excelled under external turbulence had defined, practiced, and executed their procedures with consistency. The challenge becomes balancing the need for that consistency with changing conditions. One example of how changing conditions can lead to deviating from consistent procedures is the following situation: Captain Al Hayes was piloting a United DC-10 commercial aircraft on which the tail engine exploded, making it extremely difficult to control. There was no emergency procedure for this situation, so he had to adapt and discover an alternative way to fly the plane using the throttles which controlled the engines on each wing [9]. This was not something defined in emergency procedures and not something he had trained for, but after assessing the situation, Captain Haynes was able to land the aircraft with minimal loss of life. Despite the deaths, Captain Haynes' ability to adapt is a great example of making a calculated decision with limited resources, no applicable procedure, and under stressful circumstances.

6.2.4 Implementing the Decision

After making the decision, leaders need to communicate the decision, allocate the resources, account for the time it takes to implement the decision, and re-assess how implementation of actions affects the crisis environment. The situational awareness process is continuous. Having effective communication channels will ensure rapid communication of the decision so that responders can act on implementing the decision and others can update their projection on current situational awareness. The decision process also requires ensuring that sufficient resources are available in order to implement the decision: human, time, equipment, supplies, infrastructure, procedures, trained personnel, financial, etc. The logistics and planning sections of the ICS should handle marshaling these resources. Leaders must account for the time it takes to implement an action in the decision-making process. For example, if it takes 36 hours to evacuate an area and people will be exposed to a hazard such as a hurricane during this time frame, it might make sense to implement sheltering in place. The perfect decision with insufficient time for implementation ultimately is an ineffective decision. Regardless of the decision, remember that failing to make a timely decision is in fact a decision which should be weighed against the alternatives being considered. Finally, the decision when acted upon will affect the situation and there are likely multiple decisions being made concurrently that are affecting the situation. The process does not stop after each decision is made. The situation must continually be reassessed with updated information until the crisis has been resolved through recovery.

Summary of Key Points

- Critical thinking involves skillful conceptualization and synthetization of knowledge. It must be unbiased and open to new information.
- The relationship between situational awareness, critical thinking, and decision-making is paramount to one's ability to synthesize information. Having a high degree of situational awareness partnered with the ability to acutely examine the given information allows one to make a strong decision.
- Decision theory without time constraints allows for the decision maker to gain more information to make a better decision. However, after a certain point the decision maker may suffer from information overload, creating an inability to make a good decision.
- Decision theory with time constraints puts pressure on the decision maker to make a decision quickly. If a decision is not made in a timely manner, the quality of that decision will likely be poor. Rather than spending excessive time trying to gather information, decision makers need to quickly develop an imperfect yet acceptable decision.
- The Data Quality Objective (DQO) model for decision-making framework provides logical steps that take into account inputs, criteria, and conditions that are required to make a decision. This model produces data of sufficient quality and quantity, allowing the decision maker to develop an objective solution.
- The Subjective/Empirical model for decision-making framework relies on expert experience partnered with observation of data to make informed, high-quality decisions. This model is effective when experts are involved, but it is often challenging when the data does not arrive in a timely manner or is too new to rely on.
- Implementing the decision involves communicating that decision, allocating resources, accounting for the time it takes to implement, and reassessing how the implementation of the decision will affect the crisis at hand.

Keys to Resilience

- Leaders should designate experts who are tasked with developing alternative courses of action during an incident. When experts are not available, leaders must identify means to access reach-back support from external parties.
- Response organizations should document in plans and practice the fusion of data, interpretation of data to create useful information, projection of that information, analysis and decision-making through functional exercises. Decision makers need practice in this process.
- Clearly defined roles and responsibilities within the ICS will ensure everyone knows how they contribute to supporting decision-making and risk management.
- Pre-incident, hazard-specific plans can create a specific, methodical, and consistent approach for responding in the initial phase of a crisis. These early decisions are critical to successfully and quickly moving into recovery and averting the crisis. These can be thought-out during the pre-incident phase.
- Critical thinking is a necessary skill in crisis management. While theory can be taught, the skill must be developed through practice. While most exercise practitioners strive for realistic (sometimes too predictable) exercise scenarios, crisis leaders need to be challenged with some level of uncertainty, where all information cannot be known and decisions result in somewhat imperfect outcomes. This is truly the realistic exercise. It requires a blend of creativity and technical expertise to develop challenging exercises which help develop these skills.

Application

Exercise 1: Train Collision

A train derails and catches fire. Initial reports are that it resulted from an accident with a vehicle at a crossing within your community. Your situational awareness tool shows an overview of the location with the rail line.

- What data would you like displayed in the EOC? On-scene?
- What decisions will you need to make (answer for your role in the community, IC, political leader, EOC manager, PIO, etc.)?
- What are you perceiving at this point?

Emergency Operations Center (EOC)

The EOC activates with the following situational awareness tools: multiple screens showing weather, news stations, social media feeds, and a geographic information system showing the roads, rail line, and response assets super-positioned on the map. You learn that the train was transporting Hazmat.

- Which screens are showing data/perceiving?
- Which are showing information/comprehension? What is useful?

Severe Thunderstorm

As responders take action to size up the scene, the National Weather Service issues a severe thunderstorm warning for your area over the next three hours.

- What are we projecting?
- Has some information become more useful in supporting decision-making?

The regional Hazmat response team will arrive in two hours. The rail response team will arrive in four hours.

- How does this affect your decision-making and objectives?

AskRail App

Responders used the AskRail App and identified that the 50-car train was transporting 30 cars of crude oil, 10 cars of Sodium Hydroxide, 5 cars of Chlorine, and 5 cars of nonhazardous cargo. Five of the crude-oil cars derailed, some are leaking, and one is on fire.

- What is the timeframe for decision-making related to: evacuation, offensive response, defensive actions, establishing a staging area, calling in additional resources?
- How long will it take to implement these decisions?
- What additional information do you need to know to support any of those decisions with sufficient information/confidence?
- What additional information would you like to know to support any of those decisions?

Plume/Explosion Modeling

Projection: Your community recently completed a pre-incident rail plan that includes a plume model (Figure 6.6) and explosion model (Figure 6.7) for crude oil. These are just projections but based on the planning scenario this is a worst-case scenario model.

- How would you use this information and projection to support decision-making?
- How useful/reliable is this information? Good enough to make an immediate decision/action plan?
- How does the risk affect your courses of action (evacuation, shelter, do nothing)?
- Are there other projections that would help inform decision-making?

Decision-making

A major decision is needed to protect the public. There are three primary options: evacuate, shelter in place, or do nothing.

- Divide into teams based on expertise or an alternative method to formulate a recommendation. Appoint several decision makers to evaluate the alternatives and make a decision.

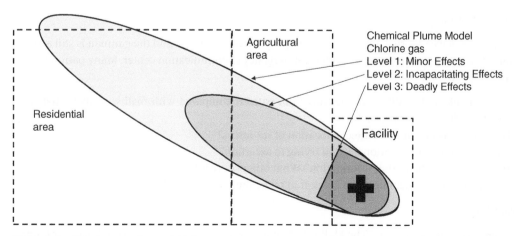

Figure 6.6 Chlorine gas plume model.

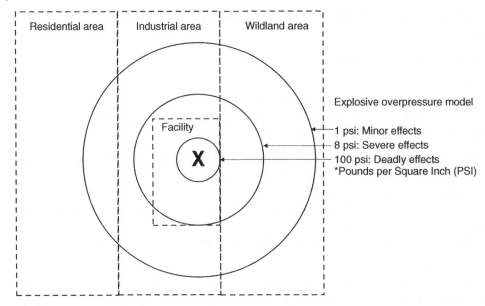

Figure 6.7 Crude oil explosive overpressure model.

- Teams should share the pros/cons of each alternative.
- Leaders should share how they evaluated the merits of the recommendation and synthesized information provided to make their decision.

Exercise 2: Active Threat

A gunman open fires in a mall. The emergency communication center is notified by patrons in the mall and in turn notifies law enforcement.

Perception

- What do you know? What do you want to know?
- What decisions need to be made?

Comprehension

Mall patrons have reported approximately 10 people have been shot and the gunman is still active. Mall security is in communication with the emergency communication center. Many patrons have safely evacuated the mall.

- How reliable and credible are the calls from patrons compared with mall security? What will you do with this information?
- How does this information affect allocation of resources?
- What additional information are you trying to ascertain?
- Where can you obtain that information? What other sources?
- How will you fuse incoming information to make the most sense of it?

Projection

Law enforcement is on scene. The Mall is locked down. There are attempts to establish communication with the gunman but there is no contact. UAVs are deployed around the mall for aerial video coverage and law enforcement has access to real-time security video from within the mall.

- What are you anticipating from this incident?
- What resources will you need to respond and recover based on these projections?
- What decisions need to be made and what information is needed?

Decision-making

- How long will it take to make these decisions?
- Formulate several alternatives to support actions needed to resolve the crisis. Compare these alternatives and discuss in groups the pros and cons of each alternative. Select the best alternative and describe your rationale for that alternative.

Exercise 3: COVID-19 Mass Vaccination

As the COVID-19 peaked at the end of 2020, a vaccine was approved. Federal and state agencies began distribution and developed plans for vaccination. This was not only a major, complex decision, but it also had bioethical elements. Competing forces included opening schools, opening workplaces, protecting the elderly, protecting at-risk populations (impoverished), protecting, and optimizing health-care worker safety, age prioritization, national security, economic recovery, mental health, and many more factors.

Work in teams based on discipline, cross-discipline, or another method. Select a crisis leader for your team. The priority objectives are: save the most lives and enable the swiftest return to normalcy for most citizens in your state or community. Applying what you have read and considered about this scenario, develop three courses of action to recommend for the appointed leader to select from and make a decision on how best to achieve the objectives. Each team should work independently in formulating distinct courses of action based on how each team is organized (i.e., discipline specific, cross-discipline, or another method). The recommendations should include some synthesis of information from various sources and may include open source information from the COVID-19 vaccination response. After the leader makes a decision based on this input, the leader should provide their rationale for the decision made. Then, discuss the pros/cons and how each alternative is evaluated against the others, and the decision-making process. Elaborate on how implementation of the decision affects situational awareness and impacts (positively or negatively) the two objectives (e.g., timeliness, metrics to monitor for effectiveness).

References

1 Scriven, M. and Paul, R. (1987). Defining critical thinking. *8th Annual International Conference on Critical Thinking and Education Reform*. http://www.criticalthinking.org/pages/defining-critical-thinking/766 (accessed May 15, 2022).

2 American Society for Quality (2022). What is the plan-do-check-act (PDCA) cycle? https://asq.org/quality-resources/pdca-cycle (accessed May 15, 2022).

3 Boyd, J. (1995). The essence of winning and losing. Five Slides. https://web.archive.org/web/20110324054054/http://www.danford.net/boyd/essence.htm (accessed May 15, 2022).

4 Boyd, J. (2010). *The Essence of Winning and Losing* (ed. C. Richards and C. Spinney). http://pogoarchives.org/m/dni/john_boyd_compendium/essence_of_winning_losing.pdf (accessed May 15, 2022).

5 University of Iowa Human Resources (2022). 8-step problem solving process. https://hr.uiowa.edu/development/organizational-development/lean/8-step-problem-solving-process (accessed May 15, 2022).

6 Burdick, B. (2016). Situational awareness: what the science says. Presented at the Virginia Emergency Management Symposium on March 31, 2016 in Newport News, VA. www.vemaweb.org/assets/docs/VEMS16/c3.pdf (accessed June 21, 2017).

7 U. S. Environmental Protection Agency (2000). Data quality objectives process for hazardous waste site investigations EPA QA/G-4HW. https://www.epa.gov/sites/default/files/2015-07/documents/g4hw-final.pdf (accessed May 15, 2022).

8 Collins, J. and Hansen, M. (2011). *Great by Choice: Uncertainty, Chaos, and Luck – Why Some Thrive despite Them All.* New York, NY: Harper Collins.

9 Kennedy, M. (2019). Al Haynes, pilot from miraculous 1989 CRASH landing, has died. NPR (26 August). https://www.npr.org/2019/08/26/754458583/al-haynes-pilot-from-miraculous-1989-crash-landing-has-died (accessed May 15, 2022).

Further Reading

1 Endsley, M. (2015). Final reflections: situation awareness models and measures. *Journal of Cognitive Engineering and Decision Making* 9 (1): 101–111. doi: 10.1177/1555343415573911.

2 The Foundation for Critical Thinking (2022). www.criticalthinking.org.

7

Application of FEMA's Community Lifelines

Decision-making within a complex, multi-agency environment can be encumbered with several challenges for private businesses, local emergency managers, state emergency operations centers, and federal agencies. Large disasters often require support from and coordination with external agencies. In order to ensure a productive working relationship and successful outcomes, emergency managers need to effectively communicate their needs and attain alignment with supporting agencies. Alignment of goals, strategies, operations, and objectives is critical to ensuring an efficient and effective response. Without proper alignment and coordination, each organization may pursue different response lines of effort according to their respective priorities and mission as each organization deems important. Crisis leaders need to avoid loss of focus and distraction from minor urgencies that arise during a crisis, and stay focused on moving the needles toward lifeline stabilization.

In the previous two chapters, we explored situational awareness and decision theory which left off at decision implementation. This chapter will explore how FEMA's Community Lifelines structure and toolkit can be used to enhance decision-making and implementation to achieve incident objectives with the support of external agencies. While attempting to strike the balance of theory and practice, I have attempted to present this material in a way that operationalizes a federally designed toolkit for our broad range of leaders from both the public and private sector. To cover the key points and objectives within the context of both the scope of this book and the intended audience, I am not able to dive into all of the details and specific examples covering every operating context. After reading this chapter, I highly recommend accessing FEMA's Community Lifeline Toolkit [1] and associated guidance documents for additional details, then working through the Application section of this chapter.

FEMA developed "the community lifelines construct to increase effectiveness in disaster operations and better position the Agency to respond to catastrophic incidents" [1]. At the time of this writing, FEMA has published version 2.0. While we anticipate that this toolkit will evolve and become more operational, this chapter will highlight several applications that can enable the crisis leader to perform more effectively by utilizing this system and toolkit. Crisis leaders need to be aware of the tools available to them; this tool can be used as a force multiplier if leaders know how to use it in accomplishing their objectives. Regardless of whether leaders adopt this specific toolkit, crisis leaders need some mechanism to organize, communicate, and direct response and recovery operations. I believe this toolkit offers the best standard for achieving this. This chapter will also provide some critique of the toolkit to help inform leaders on how best to apply and integrate the toolkit during response. Sometimes, tools like this can be overwhelming to a point of counterproductivity. But this can be avoided by focusing on the positive and actionable attributes, using the toolkit to support response and recovery, and continuously improving the tool.

Crisis-ready Leadership: Building Resilient Organizations and Communities, First Edition. Bob Campbell, PE.
© 2023 John Wiley & Sons, Inc. Published 2023 by John Wiley & Sons, Inc.

Learning Objectives

At the end of this chapter, you will be able to:

- Understand the Community Lifelines framework and utility during response and recovery operations.
- Identify the pros and cons of the toolkit from your operational perspective.
- Identify useful elements of the Community Lifelines toolkit that may lead to resource multiplication.
- Apply key elements of the Community Lifelines toolkit in the context of a disaster.
- Construct decision-support criteria within the Community Lifelines toolkit.

7.1 Terminology

In order to understand the various elements of the toolkit and get the most out of this chapter, it is important to define the terms used throughout. I am listing these definitions up-front for ease of reference. Following the definitions, I will offer a summary of how these terms relate to each other in the context of the toolkit.

Community Lifelines: the most fundamental services in the community that, when stabilized, enable all other aspects of society. Figure 7.1 shows the seven community lifelines.

Components: each lifeline is composed of multiple components that represent the general scope of services for a lifeline as shown in Figure 7.2.

Subcomponents: subdivision of components that provide granular level of enabling functions for the delivery of services to a community and help define the services that make up that lifeline [2]. For example, subcomponents of the Health and Medical lifeline, Public Health components include epidemiological surveillance, laboratory, clinical guidance, assessments/interventions/ treatments, human services, and behavioral health.

Limiting Factor: a condition that either temporarily or permanently impedes the accomplishment of a mission [2]. For example, a limiting factor inhibiting the delivery of food and water to an impacted community could be "inaccessible roads, ports, and transportation routes."

Lines of Effort: specific mission-sets required to stabilize the lifelines; operationalization of core capabilities [2].

Shortfall: Resource-specific limitations which can be determined by identifying the total requirements and subtracting available resources from the total requirement [2]. For example, a water shortfall could occur when community demand is 10 million gallons per day (MGD), but the water treatment plant can only supply 5 MGD, creating a shortfall of 5 MGD.

Figure 7.1 Community Lifelines [1].

Figure 7.2   Community Lifeline Components [1].

Stabilization: the state where critical lifeline services necessary to alleviate immediate threats to life and property are available to support the needs of survivors and responders [2].

When a disaster adversely impacts one or more components of a lifeline, crisis leaders work at the strategic, operational, and tactical levels to stabilize and ultimately restore the component(s) by establishing and implementing lines of effort. These lines of effort are comprised of intermediate objectives aimed at some end state. Shortfalls and limiting factors impede progress toward these end states, but crisis leaders leverage internal and external resources to address these gaps by communicating the component condition, impact, and resources needed to achieve the end state and stabilize the component. This toolkit provides the standardized framework for assessing the condition, communicating status, and unifying efforts toward stabilization and recovery.

7.2   Community Lifelines Construct

"The primary objective of lifelines is to ensure the delivery of critical services that alleviate immediate threats to life and property when communities are impacted by disasters. The construct organizes and aligns these critical services into one of seven lifelines which help frame the way disaster impacts are identified, assessed, and addressed" [2].

Federal, state, local, territorial, and tribal organizations deliver critical lifesaving and life-sustaining support through the delivery of 32 core capabilities spanning five mission areas: prevention, protection, mitigation, response, and recovery. Each jurisdiction defines the capacity of each capability and fields common resources which can be deployed to support a response mission. These resources are organized under the Emergency Support Function (ESF) construct. Each ESF is composed of a department or agency that has been designated as the ESF coordinator. ESFs are the "means" by which core capabilities (ways) are delivered to support stabilization or reestablishment (Ends) of a lifeline.

Since disruption to lifelines may cause threats to life and property, leaders often prioritize activities that stabilize lifelines through contingency response solutions. This is often an intermediary step toward achieving long-term recovery goals. For example, those displaced by a disaster may require temporary shelters until they can return home or to a long-term housing solution. Emergency repairs may also reestablish lifeline services within the community. As the community transitions from response to recovery, contingency response solutions are demobilized and long-term solutions are implemented, which build toward more resilient solutions.

The lifeline construct utilizes an objectives-based approach to response operations. This emphasizes the desired end state rather than the method of intervention. The objective is supported with intermediary steps to achieving the objective. For example, providing emergency temporary power may be the end state, but the intermediary steps may include deploying a team, procuring the generator, establishing the fuel supply chain, and restoring power through the generator. The status of the lifelines, components, and intermediary steps is communicated to promote unity of effort among all involved in the response operation. It standardizes terminology and criteria for assessing and reporting impacts to critical services related to health, safety, and economic security. By sharing this information in a standardized manner, responders at all levels of government increase their situational awareness of the condition, shortfalls, and next steps required to reestablish the lifeline.

When stabilization of lifelines is achieved, the community can shift its focus from response operations to recovery. The lifeline construct can continue to provide a meaningful framework for outcome-driven recovery. This model emphasizes the need for timely integration, appropriate coordination, and transparency as the community sets their own goals, coordinates resources to achieve their goals, manages their processes, and practices proper financial management to fund projects with the goal of implementing resilient recovery solutions.

While community lifelines were designed for government application, the same principles apply to private organizations impacted by a crisis. In this case, the private organization may identify different lifelines and components that align with mission-essential functions and tasks that are relevant to their organization.

The FEMA Community Lifelines Toolkit provides documents, templates, references, and communication tools. The slides and references contained in this toolkit provide a much more detailed explanation of the construct covered in this overview and should be downloaded and reviewed in more detail. Previous chapters in this book highlighted the complexity of multiagency response operations and the challenges of leading through a crisis. The remainder of this chapter will highlight the practical elements of the lifelines construct and toolkit for local and private sector-based leaders. With any tool, there are benefits and limitations associated with implementation. The next two sections of this chapter will highlight some salient points.

7.3   Benefits of the Community Lifelines Toolkit

FEMA has invested considerably in integrating this construct and tool into the National Response Framework. As stated in the FEMA Incident Stabilization Guide, this tool was designed based on challenges encountered during concurrent disasters during 2017 and therefore focuses on FEMA and strategic level application, with the expectation that other levels of government utilize this system to provide status updates, request resources, and coordinate reception and implementation of resources known as Lines of Effort. But the benefit of participation from state, local, tribal, and territorial (SLTT) agencies is that these organizations can inform FEMA and federal agencies on the status of lifelines and drive prioritization of federal support where it is needed most. In complex disasters where multiple lifelines are impacted, SLTT agencies can ensure a higher degree of situational awareness, unified effort, alignment of priorities, operational coordination, and consistency in communication.

Leaders are faced with many challenges during a crisis. While the decision theories presented in earlier chapters may work well for individual decisions, disasters tend to be very complex requiring prioritization of limited resources. There are often many competing interests and various supply chains of resources to include volunteers, commercial supplies, and equipment, as well as regional, state, and federal resources. Leaders are responsible for establishing incident objectives, compiling information, analyzing the situation, reporting the status of response operations, monitoring, and measuring progress, requesting and allocating resources, and ultimately preserving life and protecting property throughout the operation. The lifelines construct presents an organized system for articulating outcomes, aligning objectives, reporting status, and ensuring accountability. It assists with identifying severity, scale and complexity, root causes, and interdependencies of critical life-saving and life-sustaining services. When lifelines are disrupted, decisive intervention is required to stabilize the incident. This construct provides a system to support leaders in intervening decisively.

While the toolkit is robust, local leaders can customize the application of the tools based on the resources available and extent of integration with federal interagency resources. At a minimum, lifelines and their components represent assets, services, and resources that are necessary for life-saving and life-sustaining within a community. Leaders can utilize this construct to include the Lines of Effort to ensure interoperability with external resources in stabilizing lifelines so that limited resources can be applied to areas of greatest need.

7.4 Limitations with the Community Lifelines Toolkit

The lifeline tool can be elaborate, and requires specific knowledge and time to incorporate relevant information into the tool. In addition to other resource requirements during a crisis, this could seem daunting compared to making a phone call and stating the needs. For a rural community that lacks extensive personnel resources, it is incumbent on the leaders to quickly request incident management teams and personnel resources that are needed to support incident management so that local leaders can focus on setting incident objectives and broad management of the incident at the local level.

While the tool is FEMA-centric, there are elements as mentioned above that can be utilized to achieve locally driven outcomes. In some disaster situations, there is a potential for differing priorities between different levels of government and private organizations. Just as disruptions in lifelines can be communicated to other levels of government, their priorities and incident management objectives in restoring these lifelines may not align with every local entity's priorities. For example, consider COVID-19 prioritization of production and distribution of testing supplies and vaccinations. When resources were limited, there was a potential for differing priorities at each level of government.

The lifeline construct creates the potential for stove-piping problem definition and problem-solving. This should be addressed through crisis-action planning and result in a Line of Effort Operational Plan. Since there are interdependencies among the lifelines and components, it is important for ESFs to identify interdependencies when constructing the intermediate objectives and coordinating with other ESFs accordingly. For example, during the COVID-19 pandemic, a Public Health department recognized the need for establishing a mass vaccination center in suburban communities; however, the sites selected were not accessible by public transportation. Considering access needs of the population should have resulted in a limiting factor associated with limited capacity and routing for mass transit buses and trains. An alternative solution would have been to establish the mass vaccination site near public transportation routes. While Operational Level of Planning is an element of the lifeline framework, responding agencies need to recognize the conflicts and triggers to atypical solution sets.

Another limitation of the lifeline construct is that there is a potential for unresolved conflicts among the lifelines. For example, the Public Health component of the Health and Medical lifeline may be implementing intermediary actions with "lockdown" policies to reach an end state of stopping the spread of an infectious disease; meanwhile, the Economic Recovery Support Function may be attempting to implement actions that mitigate unemployment with keeping the business community open and operating. Nonetheless, the tool will highlight these diverging objectives and enable decision makers to engage in problem-solving, setting policies, and prioritizing resources.

While each stabilization target should be developed collaboratively with key stakeholders from local, state, regional, and national agencies, it is possible that these different levels of government could disagree on the stabilization target (e.g., possibly due to availability of resources, or policy disagreements). To avoid this situation, local leaders should invest in the careful development of these targets during deliberate planning even though they may be refined throughout the incident. These targets can be broadly written and refined with specific metrics and definitions as needed.

The accuracy of the lifeline condition depends on the degree of situational awareness, impact assessments, and inclusion of diverse stakeholders and partners. This is hard work and requires a whole community approach with an extensive network of stakeholders. This is best achieved during the pre-incident phase. The COVID-19 pandemic highlighted the importance of networks and collaboration between the public and private sectors in both supporting businesses to remain solvent and retain employees, and integrating business production of PPE, testing supplies, and vaccines into public sector lifeline stabilization efforts.

7.5 Application of Community Lifelines from a Local Leadership Perspective

During the pre-incident phase, emergency managers develop deliberate plans with stabilization targets across all lifelines. When an incident disrupts a lifeline, leaders must accomplish five steps in order to stabilize all lifelines. These are to:

- Assess the lifeline and component condition and designate the status of the condition. This may entail making adjustments to the stabilization targets for each lifeline.
- Establish incident priorities.
- Organize response activities around lines of effort to accomplish incident objectives.
- Establish additional logistics and resource requirements.
- Reassess lifeline conditions and status.

7.5.1 Lifeline Assessment and Status

Lifelines are assessed at the component level using six categories in order to understand what services are impacted by the disaster. This is essential information for decision makers. These categories are described according to the following method:

- **Component**: this is the component of the lifeline that is being assessed.
- **Status**: summarizes the disruption, status of the component, root causes, and changes since last assessment.
- **Impacts**: describes the extent of the disruption in terms of locations, people, and operations.
- **Actions**: describes the efforts being made to stabilize the services and address root causes.
- **Limiting factors**: identifies issues that are preventing services from being stabilized.
- **Estimated time to status change**: addresses when services are anticipated to be provided organically or through contingency response solutions.

While this is a static assessment, leaders must apply situational awareness and critical thinking as discussed in the previous chapters in order to "project" changing conditions and evaluate the rate at which conditions are changing due to the incident or implementation of decisions.

Lifeline status is communicated using a color-coded system and reflects the status of a solution to the disruption. This helps communicate priorities to decision makers.

- Grey indicates no clear understanding of the extent of the disruption and impacts.
- Red indicates lifeline services disrupted and no requirements or solution identified.
- Yellow indicates a solution identified and plan of action in progress.
- Green indicates stabilization of the lifeline.
- Blue is used only for administrative purposes and does not indicate an operational status or condition.

A high degree of situational awareness enables leaders to communicate with confidence a red, yellow, or green status. A low degree of situational awareness can be indicated with the grey status.

7.5.2 Establishing Incident Priorities

Incident priorities are derived from the pre-incident state and the need to restore conditions from the consequences of the incident back to the normal, pre-incident state. This process involves inherent risks to responders, the public, property, the environment, and the economy. Incident commanders and crisis leaders will establish incident objectives or priorities to ensure that the response actions and intermediate objectives make meaningful progress toward stabilization and restoration.

Incident stabilization is the primary strategic objective in responding to a disaster. Stabilization stops the escalation and moves the incident to a stable, steady-state condition, or one of de-escalation. However, the incident continues in some manner and conditions must be restored and reestablished. This is the overlapping bridge between the response and recovery phases. Keep in mind that recovery is a continuum that begins pre-incident and overlaps with initial response during short and intermediate term-recovery operations.

By establishing incident stabilization targets for each lifeline during the deliberate planning, pre-incident phase, leaders will already have a target that provides a basis for establishing incident objectives. This can be accomplished at the strategic, operational, and tactical levels. These targets should be validated and refined throughout the incident.

By incorporating Community Lifelines into deliberate planning products such as all-hazards plans, incident action plans, and information analysis briefs, organizations take an essential step of ensuring clear, consistent communication of priorities and response status with external organizations. Organizations can use lifeline products for senior leader briefs and situational reports. This helps to consolidate critical information up front to support decision-making and priorities.

7.5.3 Operationalizing Lifelines by Organizing Response Activities around Lines of Effort

During an incident, the planning section engages in crisis action planning which may entail refinement of the stabilization targets for lifelines. These stabilization targets inform the incident action plan (IAP) which contains incident objectives. The planning section can also formulate intermediate objectives required to achieve the end state for stabilizing the lifeline. This process is commonly accomplished through the Planning "P". The Planning "P" is the process that encompasses initial response into the incident management cycle of developing incident objectives, preparing the IAP, executing the plan, and updating the incident objectives [3]. Intermediate objectives can be coordinated with the Logistics section to determine the resources needed and schedule for delivery; this represents the *Line of Effort*.

For example, stabilization of the Energy Lifeline may include three lines of effort: temporary emergency power, power restoration, and fuel distribution.

Upon approval of the IAP, the incident commander tasks the Operations Section with implementation of the plan. Based on pre-coordination related to availability of resources, a request is made to the EOC for these resources. The local EOC coordinates with the state EOC as needed and utilizes the Senior Leader Brief to report the status of each lifeline, the limiting factors, and high-priority actions required for stabilization. Resources are mobilized, deployed, and assigned to the incident. Lines of Effort are elements of a solution to resolving lifeline instability.

Thinking in this framework enables local agencies to visualize how federal interagency capabilities can support lifeline stabilization. There are 17 standardized Lines of Effort that have been agreed on by all ten FEMA regions. These can jumpstart crisis-action planning and accelerate stabilization of lifelines. These Lines of Effort can also spread over the phases of stabilization and recovery. When federal support is provided through Lines of Effort, incident leadership can anticipate the sequence of intermediate objectives, schedule for completion, and transition into incident objectives. Delivery of standardized resources can be integrated into incident management, alleviating some stress and challenges associated with integration of capabilities from other agencies into response operations. But planners should not be limited by these predesigned Lines of Effort. During the deliberate planning phase, planners should design Lines of Effort that correspond to their risk profile and stabilize their applicable lifelines. This research, design, and coordination prior to an incident is essential for resilient recovery.

Figure 7.3 shows an example of a standardized Line of Effort which describes sequential intermediate objectives that lead to an end state. This end state supports a stabilization target and incident objectives.

7.5.4 Establish Additional Logistics and Resource Requirements

Large-scale incidents require resources beyond the local capacity. Therefore, leaders must anticipate the need to request and receive additional resources. The Logistics Section of the ICS performs an essential function for leaders by ensuring accountability and provision of the resources that are needed. Following the example pertaining to Lines of Effort, an incident commander recognizes the need for interagency resources to support the health-care infrastructure that is unable to provide patient services. Requesting this line of effort will result in the arrival and staging of health-care system support resources. The Logistics Section will support reception of these resources through identifying staging locations, personnel support requirements for lodging, food, medical, and other resources, personnel accountability and assignments, credentialing, reporting lines of communication, and demobilization. Resource typing helps define the scale and scope of these resources, the capabilities offered, and support requirements.

> **Purpose: Provide federal assistance to support healthcare infrastructure that is unable to provide patient services.**

Intermediate Objectives ***Mandatory***					End State
Mobilize and stage healthcare support resources.	Provide support for triage and patient treatment.	Resupply and conduct facility sustainment operations, including staffing.	Reassess continued need of healthcare system support resources.	Demobilize healthcare system support resources.	Healthcare delivery system is able to meet community patient care needs without the support of federal resources.

Figure 7.3 Line of Effort Example – Health-care Systems Support [1].

Leaders should anticipate the volunteers that will arrive, how to vet them, utilize these resources, and ensure unity of effort. The Logistics Section receives these resources and communicates their availability to the Planning and Operations Sections.

7.5.5 Reassess Lifeline Conditions and Status

As intermediate objectives and stabilization targets are reached, leaders must reassess the lifeline condition and status, share this information with other leaders and ensure operational coordination of resources. This step helps determine whether the Line of Effort was effective in achieving the target and desired outcome. In previous chapters, we addressed the role of leaders in monitoring performance with measures of performance and effectiveness. If the Line of Effort is effective in achieving the target, then resources shift to other Lines of Effort until the lifeline is stabilized. But if the Line of Effort is ineffective, then leaders must devise another Line of Effort to achieve the desired outcome.

Decision makers require high-fidelity situational awareness in order to make good decisions. The Senior Leadership Brief is designed to communicate lifeline component status, priorities, and resource needs. It generates the *Status (What?), Impact (So What?), Actions (Now What?), Limiting Factors (What's the Gap?), and Estimate time to status change (When?)* in a standardized brief. The templates provided in the toolkit may be useful for documenting and communicating this information to leadership to facilitate decision-making, priorities, and resource allocation.

The incident-approach template begins by highlighting the senior leader decisions needed in the next 24 hours. This table format shown in Figure 7.4 puts the most critical information up front for senior leaders and provides the criteria, conditions, risk, and follow-on actions needed.

This information is essential to maintaining a high level of situational awareness so that decisions related to employment of the right resources can be made. Leaders not only monitor instantaneous reports, but leaders need to perceive the rate of progress, and any acceleration of progress required to maintain a safe and operationally relevant effort. With many types of disasters, external conditions can change rapidly (e.g., weather, stability of structures, personnel availability, etc.). Leaders must assess the operations tempo of the response operations and continuously reassess how other factors could affect accomplishment of the incident objectives within an operationally relevant timeframe.

Line of Effort	Decisions	Criteria/ Conditions	Risks	Follow-On Actions
Temporary Emergency Power	Order additional generators	Supply deemed insufficient to accommodate requests past seven days	Power restoration may occur faster than order-to-install time	Notify logistics of order requirement

Figure 7.4 Senior Leader Decision Example [1].

Brock Long, former FEMA Administrator, understands what it is like to experience information overload. During his time as the FEMA Administrator, the US experienced hundreds of massive disasters from major hurricanes in the east to raging wildfires in the west. In an attempt to create situational awareness, Long was regularly briefed by 15 emergency support functions for each disaster and conversed with each governor that had requested support. However, this created information overload. By creating the community lifelines, Long streamlined the processing of information to prioritize information and situations requiring a decision in order to stabilize the incident with federal support. He not only demonstrated how a leader can rebuild established systems, but also how a leader can collaborate with multiple state and federal agencies in order to enhance situational awareness and decision-making.

7.5.6 Stabilization and Recovery

Once the lifeline is stabilized, efforts transition into the intermediate and long-term recovery. There may be some overlapping activities and some components may transition to recovery while other components have not yet been stabilized. As leaders make the shift to recovery, this is also the time to formulate long-term permanent solutions to enhance resilience. These are known as recovery outcomes. While the incident is fresh and stakeholders are engaged, emergency managers should solicit lessons learned, observations, and best practices for the after-action report. This input may also serve recovery efforts to not just rebuild or restore systems and infrastructure to the pre-incident condition, but rather incorporate resilience into the redesign and restoration of the systems and infrastructure that failed during the incident.

7.6 An Organizational Perspective

As mentioned in the introduction, this toolkit was designed primarily by FEMA for use by SLTT stakeholders. However, extrapolating this framework and toolkit to private organizations is worth exploring. While each organization is unique, I want to offer a framework for making this extrapolation and encourage private organizations to utilize this framework in business continuity planning and response-to-recovery operations. This planning process can enhance business resilience. Additionally, economic development organizations may also play a strategic role in collaborating with the private sector to ensure a resilient, regional economy.

7.6.1 Adapting Lifelines and Components

Organizational lifelines may be analogous to mission essential functions. For most organizations, these may include revenue generation, organizational development, operations management, financial management, and serving stakeholders. These functions are essential to continued existence of an organization. When any of these functions are disrupted, the organization loses some stability.

While each organization is unique, Figure 7.5 provides some example mission-essential tasks which may be analogous to components. Likewise, these could be broken down further into subcomponents or subtasks.

7.6.2 Detecting and Reporting Unstable Components

Organizations equipped with key performance indicators can monitor and measure performance. This may occur continuously (e.g., SCADA system) or periodically (e.g., monthly operations metrics or financial statements). When organizations identify the right indicators that align with their mission, goals, objectives, and action plans, then they are able to detect disruptions and take corrective actions. However, in a crisis situation, the disruption may occur before detection or reporting. Nonetheless, defining minimum goals and targets for each indicator can provide the "condition" associated with essential tasks. These conditions may be reported to managers, executives, board members, subcontractors, vendors, and other stakeholders. But just as with the senior leader brief, these conditions should be accompanied by the impact, actions being taken, limiting factors, and estimated time to status change or recovery. In a widespread disaster that impacts many organizations, organizational leaders may need to report and coordinate activities with external parties (e.g., power company, internet provider, landlord, economic development organizations, financial institution, law enforcement, etc.).

7.6.3 Stabilizing through Lines of Effort

When a mission-essential task is destabilized, leaders can work to stabilize and restore the task to full operating capacity. This may include personnel, procedures, systems, and financial resources. Continuity of Operations Plans (COOP) or Business Continuity Plans (BCP) should identify

Revenue Generation	Organizational Development	Operations Management	Financial Management	Serving Stakeholders
Development	Recruiting	Infrastructure	Accounting	Customer service
Marketing	Hiring	Information	Auditing	Community
Ordering	Training	systems	Tax compliance	outreach
Sales	Health/Safety	Compliance	Payroll	Employee benefits
Invoicing		Design,	Payables	
Receivables		development,		
		delivery		
		Project		
		management		
		Contract		
		management		

Figure 7.5 Organizational Lifelines and Components Model.

mission-essential tasks, procedures, resources, alternatives, and key personnel succession. By identifying the organization risks, the organization can predetermine some of the impacts that could occur and outline recovery procedures in the COOP. These recovery procedures are analogous to lines of effort which list intermediate objectives, associated resources, and end states required for stabilization. See the Further Reading section for COOP/BCP resources.

7.7 Application Summary

Applying the Community Lifeline toolkit to an organization or community may require some customization from the resources provided in the toolkit. Nevertheless, this concept is foundational in connecting situational awareness, decision-making, implementation, stabilization, and recovery. Without a framework for implementing decisions to stabilize an incident and reporting conditions and decisions to stakeholders, crisis leaders may become trapped in their organizational or community domain without adequate coordination and collaboration with external partners. Working with external partners in unity of effort will create a force multiplier and the synergy needed to stabilize the incident and work productively toward recovery outcomes.

Summary of Key Points

- There are seven lifelines comprised of components and subcomponents that represent the most fundamental services in a community that, when stabilized, enable all other aspects of society to function.
- Components and subcomponents of lifelines represent the integrated network of assets, services, and capabilities that are used day-to-day to support the recurring needs of the community.
- When disrupted, decisive intervention is required to stabilize the incident.
- Lifeline stabilization targets lead to intermediate objectives and outcomes. Intermediate objectives can be standardized with predefined Lines of Effort for interoperability with federal resources, or designed to fit the organization or community's need.
- Standardized communication tools and terminology can be used to report lifeline component status and condition in order to prioritize resources and enhance situational awareness.

Keys to Resilience

- Incorporate Community Lifelines into deliberate planning products by establishing stabilization targets for each lifeline.
- Organize Threat and Hazard Identification and Risk Assessments by lifeline.
- Assess lifeline vulnerabilities throughout the community. Reduce vulnerabilities through planning and mitigation activities.
- Customize Recovery Outcomes that emphasize long-term resilient solutions across lifelines and other aspects of the community.
- Exercise use of the Senior Leadership Brief to drive organization-wide implementation of the lifeline construct; implementing this toolkit will accelerate the transition from response to recovery, standardize communication with external agencies, and enhance situational awareness for leaders.

Application

1) What are several strengths of employing the Community Lifelines to approach pre-incident, during the incident, and post-incident during recovery?
2) What are some of the challenges associated with implementing Community Lifelines?
3) What are the several elements that when incorporated into your incident management system, would add value and improve unity of effort?
4) Download the Community Lifeline Toolkit from FEMA. Utilize these tools in accomplishing the following tasks while completing an exercise. This activity should be conducted with your team or in a group. As you work through the process, identify key decision points. You may use the following basic exercise scenario.

- Task 1: Establish a stabilization target for the applicable lifeline (e.g., Energy).
- Task 2: Describe your lifeline vulnerabilities related to this lifeline.
- Task 3: Identify the impacted components and subcomponents that your organization is focused on addressing.
- Task 4: After reading the exercise scenario, revise your stabilization target based on both a "contingency response solution" and "re-establishment of lifeline service."
- Task 5: Assess the condition and report the status of the applicable lifeline components using the categories listed in this chapter (i.e., status, impacts, actions, limiting factors, estimated time to status change).
- Task 6: Define incident priorities/objectives around unstable lifelines and components.
- Task 7: Establish intermediate objectives to reach end states that align with incident priorities.
- Task 8: Determine additional logistics and resource requirements to accomplish these objectives and stabilize the incident.

Scenario (Day 1): A derecho (major windstorm) disrupts power generation and distribution in your region. The power company normally can restore power within 24–48 hours, but due to the extensive nature of the downed power lines caused by major roadways that are blocked by trees and the short warning time for this storm, the power company is estimating 3–5 days before power can be restored. Your continuity of operations plan identifies several essential functions which require power. Thankfully, you have a backup diesel generator which can operate for 24 hours. However, the demand on gas stations has caused a shortage of fuel in the region. With so many major road blockages, fuel-truck distribution is delayed.

After completing task 8, read the update to the scenario and readdress tasks 6–8 as needed, then continue with the following questions and tasks.

Scenario Update (Day 2): You are at 50% capacity on diesel fuel and you have been only performing essential functions for the last day. Using your organization's continuity of operations plan and essential functions, devise a plan to conserve fuel and further prioritize essential functions. Second, what external resources and organizations are available to support energy requirements? If none, is there an alternative location outside of the region where essential functions can be performed?

- Task 9: Create a senior leader brief (either among your team members, or for your supervisor or senior official that needs a status update). Identify key decisions, criteria/conditions, risks, and follow on actions that are needed.

Scenario Update (Day 3): You have been able to secure a small supply of diesel fuel for your generator which will last until Day 5 when operating at 50% capacity.

- Task 10: Reassess the condition and status of your lifeline component.
- Task 11: Identify long-term permanent solutions that would improve the resilience of this lifeline.

References

1 Federal Emergency Management Agency (2020). Community lifelines implementation toolkit. https://www.fema.gov/emergency-managers/practitioners/lifelines-toolkit (accessed May 22, 2022).
2 Federal Emergency Management Agency (2019). FEMA incident stabilization guide (operational draft). https://www.fema.gov/sites/default/files/2020-05/IncidentStabilizationGuide.pdf (accessed May 22, 2022).
3 Federal Emergency Management Agency (2015). Incident action planning guide revision 1. https://www.fema.gov/sites/default/files/2020-07/Incident_Action_Planning_Guide_Revision1_august2015.pdf (accessed May 22, 2022).

Further Reading

1 Federal Emergency Management Agency (2018). Continuity guidance circular. https://www.fema.gov/sites/default/files/2020-07/Continuity-Guidance-Circular_031218.pdf (accessed May 22, 2022).
2 Federal Emergency Management Agency (2020). Community lifelines. https://www.fema.gov/emergency-managers/practitioners/lifelines (accessed May 22, 2022).
3 ISO 22301:2019 (2019). *Security and Resilience – Business Continuity Management Systems – Requirements.* Geneva: International Organization for Standardization.

Part Three

Adversity to Sound Judgment

There are several factors that create adversity to sound judgment during a crisis. These factors can be synergistic and follow the old adage, "whatever can go wrong, will go wrong." Crisis leaders need to be aware of these factors, how they impact the situation and their judgment, and what to do about them. In this section of the book, we will explore two primary factors that create adversity in sound thinking, judgment, and decision-making: stress and integration of different organizations into a unified response. Stress affects individuals and can be recognized and managed with adequate preparation and resources to overcome the stress. Coordinating, collaborating with and influencing external organizations affects the organizational response, and ability to make sound decisions. Challenges include lack of control, differing agendas, politics, and lack of conformity in capabilities (although this is getting better). In chapter 8, we will explore how stress affects judgment. In chapter 9, we will identify strategies to mitigate the impact of stress and improve judgment and decision-making.

Part Three

Adversity to Sound Judgment

8

Crisis Stress and Effect on Judgment

During a crisis, decisions made in the first minutes to hours are critical to successful damage control, prevention of death or injury, structural loss, and the overall resolution of the incident. For example, during the West Fertilizer Company incident, responders had only 19 minutes from the time of the 911 call until the explosion of the facility which stored Ammonium Nitrate. Stress undoubtedly affected decision-making.

The impact of stress on professional judgment is significant. However, the relationship between stress to judgment and decision-making is an aspect of human behavior that remains to be adequately explored. The literature in this area is extremely complex and inconclusive. But we will explore some of these human factors to improve awareness and enable leaders to create strategies to mitigate the impact of stress. There is no panacea to cure the challenges of stress while making decisions in a crisis. This chapter will focus on helping you to understand what is happening in your brain and body during stressful situations.

Learning Objectives

At the end of this chapter, you will be able to:

- Identify the psychological and physiological effects that a crisis induces on responders and leaders.
- Recognize signs, symptoms, and indicators that could adversely affect leadership decision-making.

8.1 Definitions and Spectrum of Stress

Stress occurs when one's perceived demands exceed their resources. Stress is what motivates us to get out of bed in the morning and begin our day as the perception of being too tired to awake shifts to the reasonableness of awakening which is well within our ability. Stress also increases as we consider our calendar, inbox, and to-do list for the day and wonder how we are going to accomplish everything that needs to be accomplished. It may seem daunting to begin the day with nothing completed and a growing list of demands that require prioritization, action, delegation, filing, deferring, and deleting. As we complete the tasks that are necessary (i.e., by exercising diligence), the list of demands decreases and perhaps becomes more achievable as the end of the

Crisis-ready Leadership: Building Resilient Organizations and Communities, First Edition. Bob Campbell, PE.
© 2023 John Wiley & Sons, Inc. Published 2023 by John Wiley & Sons, Inc.

day draws near. Let us consider a few other examples to illustrate normal stress, then contrast them with more stressful situations. It is important to consider the context, as one person could be highly distressed in what another person considers a normal level of stress.

Example 1: A fire department has established a three-day work shift where the workers live and work at the fire station awaiting to be dispatched for an emergency. During that three-day shift, the firefighters have a preset list of tasks, training, and maintenance that needs to be performed during this time period. Once the work is completed for the day, they are free to relax, exercise, sleep, or engage in other recreation. A relatively low-stress shift involves completion of their tasks without any emergency calls.

Example 2: During this three-day shift, the firefighters are dispatched to several small fires or other emergencies requiring their assistance. This scenario may represent a normal operations tempo for the department.

Example 3: The firefighters are dispatched to a multi-alarm warehouse fire. During their response, they learn that the warehouse stores hazardous materials and there are several neighboring residents and businesses. Unfortunately, this fire department is not trained or equipped to respond to hazardous-material incidents as they have only received awareness-level training, with a few of their senior captains having received operations-level training. This scenario represents a high stress situation where their perception of the demands is accurate (i.e., requirement for Hazmat-trained responders) and these demands exceed their resources (i.e., they do not have the capabilities needed to respond, contain, and control this incident). At this point, it is reasonable to ask, "what can they do?" and "what will they do?" It is time for crisis decision-making. If their leaders are not prepared for this stressful situation, they are vulnerable to crisis stress and poor decision-making with tragic results that may follow.

Example 4: Finally, consider a disaster incident: Superstorm Sandy formed in the Atlantic Ocean on October 22, 2012 and rapidly move up the east coast. Sandy first impacted Atlantic City, New Jersey, with winds at 90 mph. The result was a catastrophic windstorm, heavy rains, and storm surge that flooded Manhattan, tore down power lines, and caused severe damage across New Jersey, New York, Connecticut, Rhode Island, and Massachusetts. Local responders were incapable of responding due disabled resources, lack of communication and access to vulnerable areas. State responders were also impacted and overwhelmed with requests from local governments. Eight million people were without power for nine days, and ultimately it took several years to fully recover in the communities impacted by this disaster. While there were several success stories related to proactive, pre-positioning of resources and deployment of federal support, there were also lessons learned related to federal coordination and support to state and local communities. Superstorm Sandy stressed FEMA's capacity to provide effective response and recovery support. FEMA's after-action report identified "Using planning and analysis to drive operational decision-making" as an area for improving unity of effort across the federal response. Many responders did not have access to, did not know which plan to use, or did not use a plan to support operational decision-making. Using a plan during response should help facilitate decision-making and reduce stress, especially when the scope of this incident in the response and initial recovery phase entailed 1,500 decisions. Another area for improvement entailed integration of federal leader coordination and communications into response and recovery operations. E.g., "FEMA and its Federal partners also experienced challenges with accurately, clearly, and quickly communicating senior leaders' decisions to those responsible for implementing them and to those affected by them." Delays, ambiguities, or conflicts in direction from senior leaders can also increase stress for those involved in response. Finally, multiple demands on FEMA decision-makers such as the need to maintain continuity of operations in current programs and support deployments in response to the disaster created stress that caused some managers inability to make strategic decisions regarding personnel deployments [1].

While these examples describe a progressive increase in demands that most would find stressful because of the excessive perception of demands, it is possible for a seasoned leader to perform at peak level without adverse impacts from stress. Consider Olympic athletes. Many would not be able to compete, complete an event, or perform under the time pressure to compete with relevance. Yet somehow, top athletes can maintain mental toughness and calm when everything is on the line and perform flawlessly with precision. How do they do it? Why do they seem so calm and casual during the post-event interview? The interview alone would cause many people an inordinate amount of stress in front of a camera transmitting live coverage internationally to millions of people! The simple answer is that they are able to manage the stress both psychologically and physiologically. Their perception of the demands on them do not exceed their resources. They are both able to meet the demands and their mental condition enables them to know that they can handle the situation. This is peak performance under stress. This is our aimpoint for leaders experiencing crisis.

For decision makers and athletes alike, stress levels should be enough to stimulate top performance, but not enough to over-stress the body and mind. Performance declines as the body moves toward exhaustion or overload. This is an example of the inverted-U arousal-performance model of stress that is illustrated in Figure 8.1.

Since stress is affected by perception, stressful circumstances do not automatically lead to problems in judgment; rather it is the perception that leads to stress and problems with judgment. But there are other elements that contribute to the perception of the demands. These include the actual resources available, the leader's ability to marshal these resources to implement decisions, and the leader's skill in making high-quality decisions which create the conditions for successful outcomes. In the next section we will further examine the effects of stress.

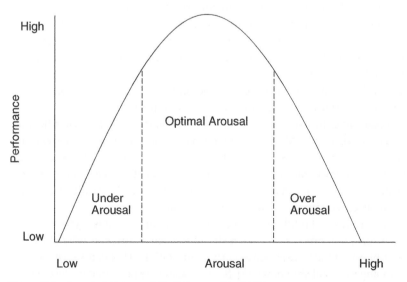

Figure 8.1 Inverted-U Arousal-Performance Model [2].

8.2 Psychological Effects

During disasters, there are often more psychological casualties than physical casualties due to the stress of the incident. Planners should project 80 psychological casualties for every 20 physical casualties. While responders and crisis leaders may be more prepared to handle a crisis than the average population, we should recognize the importance of psychological care and stress management needed for responders and decision makers during protracted crises.

Example 1: While the 1995 sarin attack on the Tokyo subway system resulted in 5,500 people seeking medical care, there were only 12 people who died, not from chemical exposure, but from heart attacks. Many of the victims attributed the signs and symptoms of anxiety and automatic arousal to intoxication from sarin.

Example 2: In 1987, when two trash scavengers in Goinana, Brazil, found a teletherapy machine and cracked open the lead and steel casing holding Cesium-137, they received a lethal dose of radiation from their exposure. Prior to their death, they sold the device to a junk dealer who was fascinated by the powder that glowed in the dark and shared it in the community. Soon afterwards, the exposed population began falling ill. Eventually, the authorities realized what had happened and directed the exposed population to evacuate into the Olympic stadium and cordoned off part of the city with no explanation. Approximately 120,000 people or 10% of the population went to the stadium due to the psychological stress associated with the risk of exposure while only 249 people were actually exposed [3].

Example 3: More recently, the COVID-19 crisis has had far-reaching psychological effects on the general population, but specifically children and young adults. During this crisis, the suicide rate has quadrupled in the US and studies have indicated a significant elevation in stress during this crisis. The American Psychological Association reported 2020 as the highest year of reported stress for adults in the United States [4], but a study done in February of 2021 projects this year's number to be higher [5]. The Kaiser Family Foundation has been tracking adult stress levels since July of 2020 and has found that more adults are reporting difficulty sleeping or eating, and increased alcohol consumption – all indicators of increases in stress levels [6]. Leaders have also exhibited signs of distress coupled with risk averseness in a highly uncertain crisis. Nevertheless, risk averseness is a major factor in the nature of governmental decision-making. Consider where you have witnessed the signs of psychological stress.

As shown in the examples above, psychological stress can present in several phases based on the level of stress. Table 8.1 below describes the progression of different types of stress.

A little stress can be a good thing. Eustress is considered beneficial stress that impacts health, motivation, or performance. However, just like an accident can cascade to a disaster, eustress can cascade to distress and post-traumatic stress which can lead to post-traumatic stress disorder (PTSD) or burnout. Therefore, it is important to recognize the early indicators that eustress is progressing into a state of distress; these are listed in Table 8.1. Ignoring these early warning signs can result in effects that ultimately cause dramatic changes in performance.

During a crisis, distress causes cortical inhibition, or not thinking clearly. Because synaptic excitation and inhibition are inseparable processes, when a stimulus, such as stress, overwhelms the brain with excitation, the brain will also experience cortical inhibition. This can range from a partial shutdown of processing to a total shut down [7]. Some researchers posit that we lose 10–12 IQ points under stress [8]. Before a crisis, during a period of eustress, our brain leads with logical

Table 8.1 Progression of Types of Stress.

Type of Stress	Eustress	Distress	Post-traumatic Stress
Leading indicators	Beneficial stress • Healthy • Motivated • Positive performance	Bad stress • Chronic fatigue • Feelings of boredom • Increasing anxiety • Vague feelings of depression	Inability to function Stress-induced reactions Unproductive
Effects	• Clear thinking • Logical processing • Balance • Thriving	• Problems in decision-making • Increased risk taking • Loss of attention • Anger, irritability • Loss of concentration • Withdrawal from others	Shutting down Indecision Mental paralysis

thoughts over emotions. However, during a crisis, we are prone to an increase in stress or even distress. During this period, our brain leads with feelings or emotions over logical thoughts. Prior to this point, our brain wants to process facts and information, but once we reach this point we benefit from support and affirmation over facts and logic. The increase in stress reduces vigilance, reduces the capacity of working memory, restricts the ability to perceive and understand dynamic conditions, causes premature closure in evaluating alternatives, and results in task shedding. These behaviors in response to distress can lead to poor decision-making, but not necessarily. Both environmental conditions and our reactions to these conditions are key elements in learning about judgment and stress.

Example: In a study, 40 subjects played a computerized fire-fighting game [9]. Half the subjects had a disturbing noise played during the game to add stress to the environment. The other participants did not experience the disturbing noise or added stress. The exercise involved varying levels of difficulty and lasted five hours. Researchers found that the stressed subjects performed equally well to the unstressed participants, but their problem-solving patterns differed. The stressed subjects focused on the general outline of the problem, while non-stressed individuals relied on in-depth analysis. Consequently, stressed subjects made fewer errors in setting priorities while the non-stressed subjects controlled their fire-fighting operations better. Adding stress to a situation may have positive effects as well as negative effects; this may lead to some peace-of-mind for decision makers.

8.3 Physiological Effects

In addition to the psychological effects of stress, the human body is designed to experience and react to stress physiologically as well. Distress affects the nervous system and can lead to a variety of signs and symptoms, commonly referred to as flight or fight reaction. These physiological signs and symptoms are created when the autonomic division of the central nervous system signals the adrenal glands to release hormones called adrenalin (epinephrine) and cortisol. This is accomplished through the sympathetic nervous system (SNS). These hormones cause the blood vessels to dilate, heart to beat faster, increase in respirations and an increase in glucose levels to enable the body to respond to the emergency or stress. Once the crisis passes, the parasympathetic nervous system (PNS) reverses the actions caused by the SNS. The reversal process causes the body to calm down by promoting bronchoconstriction and vasodilation. Both the SNS and PNS modulate stress levels but also interact with the immune system. Frequent or chronic stress may engage the SNS and PNS more frequently, which floods the body with hormones, activating and deactivating physiological responses causing wear-and-tear on the body.

Other physiological effects of stress can result in tension throughout the musculoskeletal system such as tense muscles and tension-type headaches or migraines, constriction of the airway causing shortness of breath and rapid breathing, constriction of the heart muscle, and gastrointestinal imbalances that present as "butterflies" or other discomfort which affects the bacteria in the stomach and the overall digestive process.

8.4 The Effects of Stress on Judgment

The physiological and psychological effects of stress impact competence in judgment. By definition, the decision-making process involves systematically evaluating all relevant decision alternatives before taking action. Let's examine several ways in which stress affects decision-making.

The Federal Aviation Administration identifies five hazardous attitudes that affect a pilot's decision-making. These are anti-authority, impulsivity, invulnerability, macho, and resignation. These attitudes or traits can become exacerbated under stressful conditions associated with piloting an aircraft and ultimately affect the decision-making process. Similar to managing a crisis, a pilot is constantly perceiving data from flight instruments, interpreting the meaning, identifying trends, projecting into the future as the flight process involves high speeds and changing environmental conditions which can rapidly affect the stability of flight. The pilot must maintain a high degree of situational awareness and ability to make decisions in a timely manner to maintain stable and safe flight conditions. But as stressors mount, the five attitudes which are normally in check can adversely affect decision-making.

Anti-authority may manifest itself during a crisis when time is critical, and leaders are looking for shortcuts to get the job done quicker. For example, an Unmanned Aerial Vehicle (UAV) pilot follows a checklist, conducts a safety briefing, and notifies the control tower before operating a UAV over a disaster scene. But one might rationalize, *after all we are first responders, the government, and it is an emergency.* During a crisis, leaders must still exercise responsibility within the bounds of their authority.

Impulsivity drives us to do something quickly. This is very true in an urgent crisis situation. *Just do something.* The urgency of the situation causes stress which leads us to choose alternatives more quickly before considering all of the alternatives; or we take shortcuts that we normally would not take, such as rushing into a burning building or acting on lower priority tasks. Sometimes we do need to act quickly or streamline steps in a process that is not critical to safety, but we also need to be cognizant of how the stress of the situation is affecting our ability to make sound decisions. Pausing to think, or talking out loud through the alternatives can help counter impulsivity.

During a stressful crisis, the attitude of invulnerability may present when the situation is overwhelming. *After all, how much worse can it get? I am still alive and made it through this far.* This attitude manifests in a tendency toward discounting risks and taking on greater risks. The risk-assessment process can help counter this attitude. While our risk profile may change throughout the crisis, leaders must manage risks recognizing unacceptable risks, time-based risks, and level of control over the risks.

First responders and crisis leaders may already have a high degree of confidence in their abilities based on training, experience with other crises, and position within the organization. Each time we successfully navigate a crisis, it builds confidence. This can also lead to overconfidence and a strong desire to accomplish the incident objectives. However, we must be cautious that our competence and confidence do not fool us into believing that we can accomplish something that is beyond our capabilities. The stress hormones released in our body may feed this attitude through the rush of oxygen, faster heartbeat, and increase in energy.

In extreme situations, where the situation is so overwhelming that we can no longer see viable alternatives or solutions to the problems encountered in the crisis, leaders may reach their limits and become subject to resignation. *There is nothing I can do; it won't work.* This attitude causes decision makers to give up when faced with an insurmountable task or situation in which they lack adequate control to change the endpoint of the situation. Our desire to exit the situation may cause premature selection of alternatives, shortcuts, risky and irritable behavior.

During a crisis response situation, the relative balance between our thoughts and feelings is disturbed and this can lead to risk taking and risky behavior on the part of the crisis decision-maker. As we enter the fight or flight response mode and stress hormones are released, our brain energizes the body to act. We are less able to think logically and process analytically, but more able to act, emote, and feel as the shift in hormones impacts our body.

Stress decreases our vigilance. This negatively affects our situational awareness as our brain shifts from processing and analyzing. The brain's working memory also becomes less able to contain information as the body's hormones are gearing up for a physical event. As someone presents complex information, the brain limits how much is stored and processed. Simple elements of information will help a stressed person with handling information. Instead of collecting, compiling, and analyzing all of the information, the brain selectively samples or chooses which elements of information to process; this is referred to as cue sampling. As the brain shortens the processing time and working memory cannot maintain alternatives, a stressed person prematurely discontinues collecting and evaluating alternatives and has a tendency to select the first alternative available. As working memory constricts, the stressed person will shed tasks. This can manifest in a disregard for tasks or forgetting to accomplish a task, not necessarily delegation. As the previous example study illustrated, stressed people tend to prioritize actions but exercise less detailed control over operations. The effects of crisis stress in many ways affect analysis, problem-solving, synthesis, evaluation, and decision-making.

Finally, decision makers need to be aware that acute, repetitive, and chronic stress can lead to increased risk taking and ultimately result in stress disorders. These can be combated with active steps to de-stress. Exercise is needed after a critical incident to help burn up neurotoxins and the lactic acid that lowers the pH of cells. Relaxation or breathing exercises can help our body return to pre-incident operating conditions. Journaling as little as 20 minutes a day can serve as an outlet to dump stress. Also, decision makers and their community should be vigilant about the early symptoms of distress and post-traumatic stress. Preventing stress disorders can be accomplished through early detection and intervention.

Summary of Key Points

The effects of stress include:

Psychological	Physiological
Cortical inhibition, or not thinking clearly	Fight or flight reaction
Risky behavior, increased risk-taking	Heart racing
Anger, irritability	Cold sweat
Loss of attention	Dry mouth
Problem decision-making	Digestion shut down
Loss of concentration	Elevated blood pressure
Withdrawal or resignation	Headaches

Recognize the indicators that could adversely affect leadership decision-making such as chronic fatigue, feelings of boredom, increased anxiety, vague feelings of depression.

Stress impacts decision-making in the following ways:

- Decreased vigilance
- Decreased working memory
- Decreased cue sampling
- Premature closing of alternatives
- Task shedding
- Increased risk taking
- Shift from logical thinking to an emotional state
- Changes in problem-solving patterns – broader prioritization, less in-depth analysis

Keys to Resilience

- Critical incident stress management should be initiated before a disaster to help promote wellness and health among responders and decision makers to set the conditions for successful stress management.
- Recognize that workplace stress is a workplace hazard that must be assessed for risk factors and controlled through risk management just like any other workplace hazard.
- Training, resources, and access to strategies for stress relief are necessary elements of an employee benefit package.

Application

1) Describe a situation in which you experienced distress. What were the circumstances? What psychological effects did you experience? What physiological effects did you experience? What other environmental stressors added to your stress (e.g., physical strain, heat, humidity, altitude, etc.)? How long did the stressful situation continue? How did your body react when the situation resolved? How long did it take to return to normalcy?

2) Describe a situation when you had to make important decisions while under stress. What were the decisions? How did stress impact your decision-making ability? Relate the impact to the factors listed in this lesson.

3) Experiment with a stress test. Keep it safe. For example, turn on the TV at a high volume, sit within a few feet of it. Set a timer for two minutes and attempt to write a coherent essay summarizing this chapter from memory before the timer expires. People handle stress differently by tuning out the noise or disregarding the timer. But if your grade depended on ensuring that 100% of the physiological and psychological effects mentioned in this chapter were captured, that might increase the stress. Simulators are also effective in inducing some level of stress while performing tasks. You may identify other ways to test your response under stress but keep it safe.

References

1 Federal Emergency Management Agency (2013). Hurricane Sandy FEMA After-Action Report. https://s3-us-gov-west-1.amazonaws.com/dam-production/uploads/20130726-1923-25045-7442/sandy_fema_aar.pdf (accessed May 25, 2022).

2 Anderson, K. (1990). Arousal and the inverted-U hypothesis: a critique of Neiss's "reconceptualizing arousal". *Psychological Bulletin* 107 (1): 96–100.

3 International Atomic Energy Agency (1988). *The Radiological Accident in Goiania*. Vienna: International Atomic Energy Agency.

4 American Psychological Association (2020). Stress in America 2020: stress in the time of COVID-19, volume one. https://www.apa.org/news/press/releases/stress/2020/report (accessed May 25, 2022).

5 American Psychological Association (2021). U.S. adults report highest stress level since early days of the COVID-19 pandemic. https://www.apa.org/news/press/releases/2021/02/adults-stress-pandemic (accessed May 25, 2022).

6 Kaiser Family Foundation (2021). The implications of COVID-19 for mental health and substance use. https://www.kff.org/coronavirus-covid-19/issue-brief/the-implications-of-covid-19-for-mental-health-and-substance-use (accessed May 25, 2022).

7 Isaacson, J. and Scanziani, M. (2011). How inhibition shapes cortical activity. *Neuron* 72 (2): 231–243. doi: 10.1016/j.neuron.2011.09.027.

8 Lating, J. (2017). First responder stress psychology. Presented at the Virginia Emergency Management Symposium on March 24, 2017 in Williamsburg, VA.

9 Kowalski-Trakofler, K., Vaught, C., and Scharf, T. (2003). Judgment and decision-making under stress: an overview for emergency managers. *International Journal of Emergency Management* 1 (3): 278–289.

Further Reading

1 American Psychological Association (2022). Stress index page. https://www.apa.org/topics/stress (accessed May 25, 2022).

9

Overcoming Stress to Optimize Performance

Chapter 8 covered the psychological and physiological effects of stress and how they impact decision-making. One central phenomenon is the brain and central nervous system's response to acute stress, which results in stifling the logical thinking functions while inducing hormone changes that prepare for physical stress causing emotional and physical dominance. This cortical inhibition and fight or flight response affects a leader's ability to think, process, synthesize, and evaluate information in order to make good decisions. This chapter will explore strategies for preparing crisis leaders for stressful situations so that they can prevent and overcome the effects of stress and optimize their decision-making and leadership performance.

Table 9.1 lists specific strategies for preventing and overcoming specific effects of stress identified in Chapter 8. Additionally, we will explore broad strategies for optimizing performance such as planning, organization, use of equipment, technology, and decision-support tools, training, and exercises.

Learning Objectives

After reading this chapter, you will be able to:

- Identify ways to mitigate stress and its effect on decision-making.
- Understand the role of planning in reducing crisis stress.
- Identify organizational strategies to mitigate stress during a crisis.
- Understand the pros and cons related to equipment and decision-support tools in mitigating stress during a crisis.
- Identify how realistic and stressful training and exercises condition and prepare individuals for overcoming distress.

9.1 Health and Wellness

Preparing for a crisis begins long before the incident occurs. Likewise, preparing a person mentally, physically, and emotionally for dealing with stress must begin before the stressors occur. This begins with ensuring the body, mind, and spirit are ready to handle the stressors. Because stress affects the mind's ability to think clearly as well as the body's physical condition, it is critical to ensure the body can smoothly transition among these changes. Maintaining a healthy body and mind are great starting points. As mentioned in Chapter 8, the repetitive release of stress-related

Crisis-ready Leadership: Building Resilient Organizations and Communities, First Edition. Bob Campbell, PE.
© 2023 John Wiley & Sons, Inc. Published 2023 by John Wiley & Sons, Inc.

Table 9.1   Strategies for Preventing and Overcoming the Effects of Stress.

Effects of Stress on Decision-making	Strategies for Preventing and Overcoming Effects of Stress
• Decreased vigilance • Decreased working memory • Decreased cue sampling • Premature closing of alternatives • Task shedding • Increased risk taking • Shift from logical thinking to an emotional state • Changes in problem-solving patterns – broader prioritization, less in-depth analysis	• Adequate sleep and shift duration • Visual cues, common operating picture • Repository for details, incident logs • Control the tempo, establish procedures for development of alternatives, presentations • Define roles and responsibilities, exercise delegation before becoming overwhelmed • Conduct risk assessments and compare risks of alternatives • Create an environment/setting that is calm such as offices or briefing rooms that are less chaotic than an EOC

hormones and those that counteract these hormones can take a toll on the body. Preventing the effects of stress through healthy living and wellness is an essential strategy to preparing for the stress of an incident.

As listed in Table 9.1, adequate sleep is necessary for the body to rest and return to performance the next day. Most people require seven to eight hours of sleep each night to awake fresh and ready for the challenges of the next day. This should also be taken into account when establishing operational shifts and their duration to ensure that those involved in any response can perform at peak performance with a clear mind and the attention that is needed to focus on critical thinking.

Physical fitness also plays a role in fitness for duty. Aerobic fitness increases one's endurance and enables a person to maintain a lower heart rate when under physical or emotional stress. Various risk factors could induce more serious health outcomes such as cardiac arrest when under a high level of stress. Mental and physical fitness enhance a person's ability to overcome stressors and maintain calm, clear thinking.

Managing stress with rest and exercise can also be supplemented with journaling as a therapeutic strategy. Journaling provides an outlet for releasing stress by articulating information that is weighing on one's mind with the ability to process the experience. Too often, a person encounters a stressful incident only to relive it in their mind because they have not expressed or shared the experience. Journaling is a private way to allow these thoughts to be expressed and processed. This often results in relieving the stress.

9.2   Planning

"Plans are worthless, but planning is everything" reflects a common attitude toward plans and planning. This is a very important distinction because when you are planning for an emergency you must remember one thing: the very definition of "emergency" is that it is unexpected. The event is not going to unfold the way you have planned.

> "The details of a plan which was designed years in advance are often incorrect, but the planning process demands the thorough exploration of options and contingencies. The knowledge gained during this probing is crucial to the selection of appropriate actions as future events unfold." President Dwight D. Eisenhower (1957) [1].

Planning is a process that brings stakeholders together to think through various incidents, hazards, and threats, and to document information so that a group of people can prepare for, mitigate, prevent, respond to, and recover from the incident. In the last decade, the concept of "all-hazards planning" has driven planners into forging consistency and standardization in an organization's approach to incidents. FEMA published the Comprehensive Preparedness Guide 101 in order to standardize the process of planning and building maturity and consistency into community plans. FEMA also published a Comprehensive Preparedness Guide 201 to establish a process for conducting a threat and hazard identification risk-assessment process. The 2019 Emergency Management Standard established a best practice for emergency management programs that is based first on conducting a Hazard Identification and Risk Assessment as the basis for various plans. The Standard has multiple elements dedicated to establishing criteria for plans such as Emergency Operations Plans, Hazard Mitigation Plans, and Continuity of Operations Plans.

While there are many types of plans, most plans will include the following elements: scope, purpose, authorities, responsibilities, references, concept of operation/context, assumptions, procedures, communications, logistics support, resources, and plan maintenance. Some plans are shorter in nature, such as pre-incident plans, which are focused on specific facilities and situations. When done correctly, both the planning process and the plans can in fact serve a worthwhile purpose. The planning process has intangible value since it creates opportunities for trust and team building among stakeholders. It often results in sharing information and resources that otherwise are kept in a stovepipe as an institutional secret. At best, planning results in collaboration and ideation that solves problems and creates innovative approaches to addressing threats and hazards. If the result of the planning process is a plan that is developed solely to meet a requirement, then the plan can be worthless. But if the plan is a result of stakeholder collaboration that involves problem solving, resource integration, information sharing, operational coordination, and critical thinking, then the plan is worth publishing and executing. Unfortunately, many practitioners dismiss the value of the plan due to these missed opportunities only to learn or re-learn the lesson that their plan was ineffective, so they dismiss the value of the plan. Rather, we need to invest the time and collaboration needed during the planning process to develop an effective plan that helps stakeholders and leaders successfully navigate a crisis. Our attitude about plans is a reflection of the plan's effectiveness.

Plan effectiveness is a function of usefulness. Too many plans take on a voluminous nature as to render them cumbersome and useless during an incident. Therefore, it is important to develop plans that are concise, well organized, and easy to navigate, especially if it is a comprehensive plan. Some plans, such as pre-incident plans can be as short as one to two pages and contain critical information about a facility with schematic diagrams. As emergency management becomes digitized, much of this information can be contained in common operating pictures, geographic information systems, searchable libraries, workflow assistants, and QR-code labels on buildings and vehicles. Effective planning that involves incident analysis can provide decision makers with alternatives, analysis results, and recommendations which can facilitate quick decision-making during an incident. Researchers Kelly and Cool point out, the more knowledge one has about a topic, the faster one can review and process the information [2].

Three essential elements of a plan that can reduce stress prior to and during an incident are (1) clearly defined roles and responsibilities, (2) verified procedures, and (3) risk-analysis. There is nothing worse than trying to problem solve, identify responsible parties, develop procedures, and conduct a risk assessment and impact analysis during an incident. The stress of these activities is preventable with planning.

Roles and responsibilities should be clearly defined in operational plans. Other plans should ensure consistency with operational plans by referencing roles and responsibilities and then amending with specifics that are necessary for functional or hazard-specific plans. Every party needs to know their role and responsibility prior to the incident. Implementing a consistent incident command system also ensures operational coordination among each party once their role is defined.

Documented and verified procedures enable responders to quickly react under stress. As previously mentioned, stressful events cause a fight or flight reaction and cortical inhibition which shifts one's brain processing from thinking logically to thinking emotionally. Knowing this vulnerability to stress, we should minimize the amount of thinking that responders and decision makers undergo, especially in the initial phase of response as their analytical bandwidth is reduced. The best way to prepare for this situation is to follow a procedure. However, responders will only follow the procedure if they trust the procedure and it is documented. Responders trust procedures that have been verified through practice, training, exercises, and real-world experience. Responders are more likely to trust their previous experience which could vary greatly among their team based on volume and situations. This is a recipe for conflict during the response. Therefore, it is best to capture stakeholder best practices, test them, and once verified, incorporate them into procedures. Ensuring everyone on the team knows, trusts, and implements the procedures enables teams to successfully execute a response to an incident.

Procedures can entail checklists or processes that involve physical activities, actions, or decisions. Response procedures should be organized consistently across the organization and follow some standardization among the first actions to gain momentum in taking response actions. While it is very unlikely that a crisis will play out exactly as a plan "assumes," many of the first steps of any crisis will be similar. Having a well-thought-out initial response checklist helps ensure that the first steps of a response will pave the way for effective support-to-recovery actions. Trusting the procedure enables responders to act and avoid the trap of thinking when processing speeds are slow. Procedures provide a crisis leader with the ability to be decisive while clearing their mind for addressing the unexpected situations that require critical thinking.

Several regulations (e.g., Clean Air Act and Emergency Planning Community Right-to-Know Act), the Joint Commission International, and the 2019 Emergency Management Standard require organizations to identify hazards and assess the risks of these hazards in order to inform the planning process. This is known as risk-based planning. Incidents can vary in likelihood of occurrence, severity and impact on an organization and community. Since we operate in an environment where resources are constrained, a risk-based approach to planning ensures that decision makers can allocate limited resources according to risk. The environmental regulations listed earlier require organizations (public and private sector) to identify hazardous chemicals that could impact the environments and the surrounding public. Similarly, hospitals accredited by the Joint Commission International are required to conduct hazard vulnerability assessments which encompass community and hospital-specific hazards that could impact the operation of the hospital. Under the Emergency Management Standard, organizations are also required to complete a hazard identification, risk assessment, and consequence analysis.

Under each regulation and standard, an organization (the facility, public safety officials, or both) are responsible for conducting a risk assessment. A risk assessment is based on the likelihood of occurrence, severity of the incident, and the consequences. Estimating the likelihood of a chemical release may be based on industry statistics, facility-specific incidents, natural disaster hazards, and man-made activities to establish the likelihood of occurrence. The severity analysis might include modeling likely and worst-case hazard scenarios to determine the severity of the incident and the subsequent impact on people, the environment, the economy, and infrastructure. For example, a chemical release may be modeled to assess the impact on human health in downwind locations. An infrastructure analysis could assess the impact of different magnitudes of earthquakes on buildings, utilities, and transportation systems. There are multiple risk methodologies to determine risk, but the most rudimentary method is illustrated in the risk matrix that accounts for likelihood and severity. Ranking hazards based on risk enables planners and decision makers to establish priorities and allocate resources based on the greatest risk reduction, not necessarily a specific function or hazard.

Example: The NFPA 1620 Standard on Pre-incident planning establishes a framework for conducting planning at facilities in order to identify various hazards, resources, demographics, infrastructure, and fire suppression systems at the facility so that if an incident occurs, responders have a baseline of familiarity with the facility and can respond expeditiously. A pre-incident plan documents the main hazards such as bulk chemicals stored on site and their location, the access points, fire hydrants, and location of suppression systems, the type of building construction and materials, and resources available onsite. The planning process entails a physical survey of the facility which includes responders, facility managers, and operators. While surveying the facility, a schematic diagram or sketch of the facility is annotated with key information. After the survey, planners may conduct further analysis on the facility such as plume or explosive modeling for any chemical releases, determination of evacuation points for workers and the community, and establishing alternative access points and staging areas. In addition to this analysis, planners could include critical first steps and actions that initial responders could take, such as shutting off a valve, or evacuating a downwind neighborhood. This is an example of a pre-incident action plan. Pre-incident action plans enable initial responders to make swift and accurate decisions based on preplanning and analysis. These plans can also form the basis of joint training and exercises between the private and public sector responders.

9.3   Organization

The human brain can process about five to seven inputs while multitasking. The myth of multitasking tells us that a person may be able to increase their productivity when doing two tasks concurrently, depending on the task, but as a person multitasks further, their brain continues to shift their focus among each input or task. Information overload occurs when a decision maker attempts to juggle more than five to seven inputs at a given time. Also, there is the fundamental characteristic that decision makers tend to seek more information than is necessary and this information overload decreases decision-making performance. This may be evident in emergency operation centers, dashboards, and common operating pictures where information quantity abounds. In many cases, more input and information are not better; in fact, more input and information can become counterproductive and distracting.

The incident command system establishes a limit on the span of control to five to seven people reporting to another person. Again, this similar range on span of control is based on realistic limitations of what a person can handle. This establishes an organization structure that optimizes information flow, inputs, and decision-making within the span of one's control.

Case Study

During the West Fertilizer Company incident, Frank Patterson reported that the ICS was not fully established at the onset and information overload became a distraction from focusing on response actions and decision-making [3]. Several positions within the ICS were not established, and as volunteer responders created traffic jams on the highway, the responders lacked a system to handle the large influx of volunteers, process, and credentials while assigning volunteers to meaningful roles in the response. Additionally, phone calls from multiple federal agencies overwhelmed the incident commander. In one case, he hung up and handed his phone to an assistant, only to find that the federal agencies would soon show up in person to investigate the accident from a criminal and accident perspective while he was leading response and recovery efforts. Communication needs among the responders created demand for briefings and the lack of liaison officers resulted in one hospital going public criticizing decision-making which did not align with their mass casualty plan. Due to impending tornado warnings, the incident commander deviated from the plan on air evacuation to move critical casualties by air a shorter distance in order to air evacuate more casualties prior to arrival of the squall line. Since the reasoning behind this decision was unknown to the second hospital, confusion erupted. The speed and intensity with which critical decisions need to be made during a crisis can easily overwhelm a decision maker and the incident command if issues such as span of control, operational periods, and assignment to roles in the ICS are not addressed early in the response.

Crisis leaders need a streamlined and surgical method for receiving actionable information in a timely manner in order to make good quality decisions. One technique to manage information overload is to have those reporting to the leader provide summaries or dashboards rather than provide all the data that they have collected. The bottom-line-up-front (BLUF) approach is recommended – get to the point immediately and add details as needed to support presentation of recommendations or alternatives. Leaders should appoint those to key roles that have adequate technical knowledge of the area they are responsible for, the skill to be concise and decisive in weeding out superfluous information, and the ability to synthesize information quickly to formulate recommendations. Without these critical attributes, decision makers can easily become overwhelmed in solving problems for their reporting officials. Since many of those reading this fill a role where they report to a crisis leader, it is important to refine your knowledge, skills, and abilities to become effective. As you take on a leadership role, surround yourself with those that exhibit these characteristics and invest in training those around you through formal professional development and exercises. These are keys to creating a decision-making system that works.

In summary, several keys to preventing information overload include: limit the number of reporting officials and information inputs to five to seven, resist the temptation for more information, utilize summaries and dashboards limited to the inputs needed to support incident objective decision-making, and insist on the BLUF approach. As mentioned in the previous section on planning, effective plans such as pre-incident action plans can also preempt information overload.

Case Study

A study of military commanders found that teams with records of superior performance had one common critical characteristic: they were extremely adaptive to varying demands [4]. The teams in the study could maintain performance in less time than average teams. It took them one-third the time to make decisions when compared with a typical team. Researchers noticed that their mode of communication changed over the course of an event.

Initially the team responded to explicit requests for information from commanders. As time pressure increased, they stopped waiting for requests and instead provided commanders with information they implicitly determined would be useful. This is because practice is a major key to improvement of response. Hence, it is very important for responders and decision makers to train and exercise under stress, so that when disaster strikes, they are ready and able to obtain and provide implicitly useful information.

In summary, successful teams know the system. They know their role, responsibility, who they report to, what information is needed, and when to report. This record of superior performance is enhanced with frequent scenario-based training.

9.4 Equipment, Technology, and Decision-Support Tools

Decision-support tools play a valuable role in crisis decision-making. No single individual can acquire and process the diverse and rapidly expanding information needed to create and execute plans effectively. Tools such as GIS and common operating pictures can transform raw data into a picture that enhances situational awareness and improves risk management. This does not replace the role of subject-matter experts and their judgment. Keep in mind some of the limitations of these systems, which include displaying only the data that is fed into the system causing critical information gaps, time delay from when something happens to when it is displayed, and disparity among resources displayed. While these factors should not eliminate these systems from informing decisions, leaders should be careful when using these decision-support tools to ensure they are not over-relying on the information displayed. These tools tend to decrease the perception of dynamism and uncertainty under high stress conditions. In other words, these tools have the potential to mitigate the perception of stress, but they can also induce higher stress levels. Additionally, they do not mitigate the perception of information overload. While these systems can be valuable tools during a crisis, understanding the limitations and effects prior to the incident can ensure appropriate utilization.

There are many other types of equipment, technology, and decision-support tools common to emergency management. These should be tested in various conditions and under conditions of stress – time can be one of these stressors – to help evaluate their effectiveness. One useful element of many common operating pictures is the ability to timestamp log entries, activities, and decisions. This is helpful because real-time events, information, and decisions can be shared across the team. Keeping everyone informed in real time enhances situational awareness and can improve decision-making quality.

Finally, infrastructure or setting can be an important factor in reducing stress. The hustle and bustle of a fully operational Emergency Operations Center or field operational environment could detract from the calm needed for some leaders to hear, understand, process, and evaluate alternatives. The infrastructure established to support an incident should align with the organizational structure, size, and complexity. Creating an environment conducive to the right level of information sharing, coordination, and communication among team members at different levels within the organizational structure should be an important pre-incident planning consideration.

9.5 Training and Exercises

There are many forms of training: basic, qualification, competency, refresher, etc. Training is designed based on learning objectives and topics. Most of the training offered for responders and leaders is at the knowledge level. Blooms taxonomy, illustrated herewith, is utilized by instructional system designers to create learning objectives that align with desired outcomes related to the training. These outcomes can be summarized as knowledge, application, and judgment. Knowledge-based training is useful when a person requires the grammar of a subject, i.e., the ability to become familiar with terminology, systems, or procedures. This is typically verified through written multiple-choice tests to demonstrate that the participant is able to describe, recall, or define specific knowledge acquired in the training.

The second level of training is application-based training. This requires participants to demonstrate, utilize, apply, or perform a task. They not only need the knowledge of the procedure but they need to know how to perform the procedure successfully. This is typically verified through demonstration, e.g., donning and doffing PPE properly, operating technology, or physically setting up infrastructure.

The third level of training is judgment-based training. This may also be referred to as scenario-based training. This is the culmination of learning knowledge, demonstrating competency through application, and synthesizing information for evaluation and decision-making.

Table 9.2 crosswalks the cognitive process and knowledge dimensions. The cognitive process dimension expands the three levels of training outlined earlier into six categories of thinking skills. The knowledge dimension lists four categories that span the concrete–abstract knowledge spectrum. While most basic training is represented in the upper left corner, more advanced training

Table 9.2 The Cognitive and Knowledge Dimensions of Learning.

	The Cognitive Process Dimension					
The Knowledge dimension	1. Remember	2. Understand	3. Apply	4. Analyze	5. Evaluate	6. Create
A. Factual knowledge						
B. Conceptual knowledge						
C. Procedural knowledge						
D. Meta-cognitive knowledge						

that requires judgment and execution falls into the lower right section of the matrix. Crisis leaders should focus on training that stresses their knowledge and cognitive processes in order to prepare for actual events.

An analysis of recent chemical accident investigations conducted by the US Chemical Safety Board showed that while many responders had successfully completed knowledge-based training required to fill the roles performed in the ICS, they had not completed training at the application or judgment level [5]. For example, responders to the West Fertilizer Company incident had received certificates for completing independent study courses on incident command system, national incident management system, and hazardous material awareness, which are all knowledge-level training. But they had not completed application-level training such as: hazardous material operations or technician training, which would have been required to engage in defensive or offensive hazardous material response, respectively.

Case Study

Over the last 18 years, my company (Alliance Solutions Group, Inc.) has engaged heavily in designing, developing, and conducting thousands of realistic, scenario-based training courses for responders, military, industry, and hospitals. Our process starts with knowledge-based training, followed by application-based training, and culminates with scenario-based training that requires synthesis, evaluation, and decision-making. Through this hands-on process, participants increase their proficiency in responding to complex incidents, such as: mass casualty incidents, hazardous material incidents, and military combat situations. Feedback from tens of thousands of participants overwhelmingly favored realistic, scenario-based training because it is fun, relevant, builds muscle memory, and builds confidence in team members' ability to face a crisis and resolve it successfully. This type of training inherently incorporates physical and mental stress on people, plans, and resources due to the time and resource constraints.

Exercises can also incorporate the same stressors. Training and exercises may overlap in practical execution. Yet, according to current definitions, exercises are designed to test and evaluate specific objectives, not necessarily train participants. Exercises may be incorporated into training, but the traditional exercise is focused on testing and evaluation of capabilities. Exercises can be conducted in discussion-based or operations-based formats. Similar to great training, a great exercise is well-designed and conducted around objectives, evaluated on specific criteria, and focused on systems such as plans, procedures, training, and outcomes, rather than individuals. To be effective, exercises should stress participants and systems. Performing under stress has the added value of developing muscle memory to overcome cortical inhibition and reverse the effects of mental distress for real-world incidents.

9.6 Expertise, Competency, and Proficiency

Consider the following two case studies developed by Hammond on how decision makers followed different paths but achieved successful outcomes [6].

Case 1: In 1988, the USS Samuel B. Roberts, operating in the Arabian Gulf, struck a mine, caught fire, and began to sink. The Roberts' captain, Commander P.X. Rinn, drawing upon his training and experience, analyzed the situation and determined a course of action directly opposed to

Navy protocol. From his knowledge of how much water the ship could take on and still stay afloat, Rinn realized that the Roberts would sink before his crew could extinguish the fire. Commander Rinn made the decision to focus on keeping the ship afloat and give the fire second priority. He is on record as having arrived at his decision analytically, based on available information, training, and operational experience.

Case 2: A United DC-10, on its way from Denver to Chicago, lost its hydraulic fluid and hence, its controllability. Captain Al Hayes and his crew had to discover an alternative way to fly the plane by using the throttles – something their training had not prepared them for – and do it with few of the normally available cues. That they were able to land with minimal loss of life may be attributed to intuitive decision-making under stress.

"The two cases portray decision-making in life-threatening circumstances under two distinct scenarios: one where the knowledge and training of the decision maker were readily applicable and one where the decision makers' training and checklists had not prepared them for the exigency they faced. Yet, both instances involved individuals who were highly trained and had conducted extensive planning even though their specific cases were not in the plans or checklists" [4].

For these decision makers and others, becoming knowledgeable of the operating environment and available resources before the crisis, creating and training on plans and procedures under stressful conditions, relying on decision-support tools, and minimizing information overload are all the foundations of combating the negative effects of stress during disaster situations. Even though they did not rely on a prepared plan or procedure to avert a disaster, they were able to think and make decisions under the stress of the crisis. Their knowledge, proficiency, planning, and training enabled them to accomplish this.

Summary of Key Points

- Although one can never be fully prepared for an incident, having a pre-incident plan can drastically reduce stress when the incident occurs. When plans are already in place, it is easier for responders to make faster and higher quality decisions based on the available options. Reliance on plans, procedures, and checklists can reduce the risk of poor decision-making under stress.
- Having a pre-established plan enables leaders to execute plans rather than become overwhelmed by analyzing all possible outcomes and develop plans.
- Good quality decisions made during response can lead to faster stabilization and timely transition to recovery.
- Realistic and stressful training and exercises based on actual plans, resources, and operating context will build muscle memory which will help reduce the levels of mental distress when disaster strikes.

Keys to Resilience

- A crisis leader needs to be physically, emotionally, and mentally fit in order to be resilient throughout a crisis.
- A crisis leader should have demonstrated that they can successfully deal with stressful situations and maintain sound mental abilities throughout, prior to assuming a leadership role in a crisis.

- Leaders should be familiar with their organization's plans, procedures, infrastructure, technology, and resources prior to an incident. This can be verified during periodic exercises.
- Pre-incident action plans can reduce stress and enable swift and effective decisions that allow for a transition from response to recovery.
- Realistic, scenario-based training and exercises are an essential component of every organization's emergency management program as they ensure proficiency among responders and leaders, and verify capabilities prior to an incident. Knowledge of an organization's capabilities and their limitations are essential boundaries that every leader should understand prior to a crisis.

Application

1) Complete the referenced Wellness Compass Self-Assessment to evaluate your balance and readiness across eight different domains of your life [7].
2) Describe a plan that you developed and write an evaluation of the planning process and the resulting plan. Some criteria to consider includes stakeholder representation, stakeholder participation, stakeholder coordination, and organization of the planning process (schedule, roles, responsibilities, accountability). Download the 2019 Emergency Management Standard and compare your process and plan to this benchmark.
   a) What factors influenced the effectiveness of the plan?
   b) What could be done differently to improve the effectiveness of the plan?
3) Evaluate the infrastructure and organizational structure used in your organization to conduct incident response operations. How does the setup and assigned locations affect stress of the leadership team?
4) Thinking back on the last year of training, how well did this training prepare you to face a crisis situation? What aspects of the training prepared you best?
5) Compare and contrast knowledge-based training with scenario-based training.
6) Describe the last exercise that you participated in. Evaluate how well the exercise verified readiness for a stressful incident. What changes could enhance the exercise design to create more stress and test preparedness to deal with crisis stress?

References

1 Eisenhower, D. (1958). Public papers of the Presidents of the United States, Dwight D. Eisenhower, Containing the public messages, speeches, and statements of the president, Remarks at the National Defense Executive Reserve Conference (November 14, 1957), 817–818. Washington, DC: Federal Register Division, National Archives and Records Service, General Services Administration.

2 Lating, J. (2017). First responder stress psychology. Presented at the Virginia Emergency Management Symposium on March 24, 2017 in Williamsburg, VA.

3 Patterson, F. (2016). The West Texas Explosion that Rocked the Hazmat Community. Presented at the National Association of SARA Title III Program Officials Workshop on May 25, 2016 in Omaha, NE.

4 Kowalski-Trakofler, K., Vaught, C., and Scharf, T. (2003). Judgment and decision-making under stress: an overview for emergency managers. *International Journal of Emergency Management* 1 (3): 278–289.

5 U.S. Chemical Safety Board (2016). Emergency planning and response. http://www.csb.gov/recommendations/emergency-response- (accessed May 26, 2022).

6 Hammond, K. (2000). *Judgments Under Stress*. New York: Oxford University Press Inc.

7 Samaritan Family Wellness (2021). Adult wellness self-assessment. https://static1.squarespace.com/static/5b54fe275ffd2051be834f8c/t/5c5c9e4d4e17b603e1c1de1b/1549573716555/Adult+Self+Assessment.pdf (accessed May 26, 2022).

Further Reading

1 Federal Emergency Management Agency (2021). Developing and maintaining emergency operations plans: comprehensive preparedness guide 101, version 3.0. https://www.fema.gov/sites/default/files/documents/fema_cpg-101-v3-developing-maintaining-eops.pdf (accessed May 26, 2022).

2 Federal Emergency Management Agency (2018). Threat and Hazard Identification and Risk Assessment (THIRA) and Stakeholder Preparedness Review (SPR) guide: comprehensive preparedness guide 201, 3rd Edition. https://www.fema.gov/sites/default/files/2020-04/CPG201Final20180525.pdf (accessed May 26, 2022).

3 ANSI/EMAP EMS 5-2019 (2019). *Emergency management standard*. Falls Church: Emergency Management Accreditation Program. www.emap.org (accessed October 2, 2021).

4 US Department of Health and Human Services (2005). A guide to managing stress in crisis response professions. DHHS Pub. No. SMA 4113. Rockville, MD: Center for Mental Health Services, Substance Abuse and Mental Health Services Administration. https://www.eird.org/isdr-biblio/PDF/A%20guide%20to%20managing%20stress.pdf (accessed May 26, 2022).

5 Samaritan Family Wellness website. www.samaritanfamilywellness.org.

Part Four

Crisis Leadership

Previous chapters provided knowledge, theory, best practices, and case studies related to crisis leadership. This next section of the book explores insight from modern crisis leaders on real-world crises that they faced. Chapter 10 contains a series of profiles from leaders representing diverse contexts to include the private sector, local and state incident command, military, and federal government. These profiles reveal how these leaders navigated different types of crises and illustrate how they led and applied topics discussed in previous chapters. Chapter 11 summarizes the attributes and essential skills of a crisis leader based on commonalities and trends identified among the profiled leaders. While the theory and objective content are important to understanding crisis leadership, these next two chapters integrate the soft skills and subjective components that complete the portrait of a crisis-ready leader.

Part Four

Crisis Leadership

10

Profiles in Crisis Leadership

Chapter 10 puts the reader in the copilot's seat as our profiled crisis leaders provide details on how they successfully navigated a major crisis. Each profile is based on an interview conducted by the author and written in narrative form introducing the leader, the crisis, and the techniques they used to enhance their situational awareness, make decisions, and exercise influence to shape the conditions for success. During the interview, our crisis leaders shared about the stressor that affected them and how they managed stress so that they could make sound decisions. Each profile ends with their reflection on the traits and skills that enabled them to lead successfully during the crisis and their recommendations for future leaders. Review the Application questions after reading each profile.

Lesson Objectives

After reading this chapter, you will be able to:

- Recognize common crisis leadership principles, traits, and skills across diverse crisis situations.
- Identify key decision points and how the leader gained access to and utilized information to support decision-making.
- Identify stress management techniques used by crisis leaders and infer something about the effectiveness of those techniques.
- Describe how the media, political leaders, and external parties influenced the decision-making process.
- Assess how well the crisis leaders handled external influences and note particularly effective techniques.

10.1   Crisis Leader Profiles

I have selected several crisis leaders representing diverse roles and experiences in the private sector, military, and public sector at the local, state, and federal levels of the US government. Additionally, there is some diversity in the type and duration of the crises presented to provide readers with these different perspectives. One common remark from our crisis leaders was the encouragement for future leaders to read and become familiar with history, lessons learned, and the experiences of others. I hope this exhortation resonates with our readers even though some of our readers may be working in different contexts and encounter very different situations than

Crisis-ready Leadership: Building Resilient Organizations and Communities, First Edition. Bob Campbell, PE.
© 2023 John Wiley & Sons, Inc. Published 2023 by John Wiley & Sons, Inc.

those presented in this chapter. There are many lessons we can learn from studying these profiles within their respective contexts.

Here is a brief summary of our crisis leaders and what they shared during an interview:

Brock Long, Chairman of Haggerty Consulting and former FEMA administrator from 2017 to 2019, describes his experience leading FEMA through the most costly and disastrous hurricane and wildfire season, and how he refocused FEMA's strategy to foster a culture of preparedness throughout the US.

Lieutenant General (retired) H.R. McMaster, Senior Fellow at the Hoover Institution, former National Security Advisor, and 36-year veteran of the US Army, describes his experience leading the 3rd Cavalry Regiment in Operation Restoring Rights, a pivotal counterinsurgency operation against Al Qaeda in Tal Afar, Iraq in the Fall of 2005.

Frank Patterson, Emergency Management Coordinator and Risk Manager at McLennan Community College, describes his experience as the incident commander in the aftermath of the devastating West Fertilizer Company explosion in April 2013.

Derrick Vick, President of Freedom Industries in Rocky Mount, NC, shares his experience leading Freedom Industries through a major business crisis that almost bankrupted the company.

Chad Hawkins, Assistant Chief Deputy at the Oregon State Fire Marshall's Office, shares his experience leading an Incident Management Team to assist Marion County, Oregon in the early stages of response to COVID-19 pandemic.

Major General (retired) Dana Pittard, Vice President of Allison Transmission, and former Joint Forces Land Component Commander in Iraq, shares his experience leading US and Iraqi forces in turning the tide in a crucial 2014 campaign against the Islamic State in Iraq.

10.2 Brock Long: Leading FEMA through Transformation while Supporting Federal Response to Hundreds of Disasters during the Most Extensive and Costly Disaster Season in US History

The following is based on an interview that I conducted with Brock Long on April 27, 2021.

Brock Long was nominated by President Trump as the tenth Administrator of the Federal Emergency Management Agency (FEMA) on April 28, 2017, and subsequently confirmed by the senate on June 20, 2017 (prior to 2007, the position title was "Director"). He served approximately two years in the position and managed more than 220 Stafford Act declared disasters in addition to leading an agency of 21,000 personnel. FEMA's responsibility also extends to executive branch continuity of operations, countering cybersecurity, and preparing for external threats such as adversarial missile launches. While the other crisis leader profiles focus on a single incident or short campaign, Brock Long led FEMA through a multitude of major crises over a two-year period. Therefore, this profile encompasses how he responded as a leader throughout his term.

The 2017–2018 disaster recovery efforts cost taxpayers more than what had been spent over the previous nine FEMA administrators combined [1]. Hurricanes Harvey, Irma, and Maria were some of the most costly disasters to impact the US. Meanwhile, California experienced the worst wildfires in its history. At the time of Hurricane Harvey's impact, the majority of FEMA personnel were over-extended on deployments to other disasters throughout the US and had to be recalled and redeployed for this major disaster. Two weeks later, Hurricane Irma (the third most intense Atlantic storm on record) made landfall in Florida as a Category 4 Hurricane. The storm was responsible for 134 deaths and over $77 billion in damage. Throughout this chaos, smaller disasters continued to create demand on FEMA resources. Due to these back-to-back disasters, FEMA was at a breaking point. Then, two weeks after Hurricane Irma, Hurricane Maria impacted Puerto Rico as a Category 4 storm becoming the third most-costly hurricane on record.

During 2018, Hurricanes Michael and Florence impacted the US. Due to time difference and news cycles, many Americans were not aware that Typhoon Yutu impacted Saipan in the Northern Mariana Islands. This was the second strongest Category 5-equivalent super typhoon to ever impact the United States and its territories.

Brock inherited an organization that had a culture of being the nation's 9-1-1. In the aftermath of Hurricane Katrina, Congress passed the Post-Katrina Emergency Management Reform Act (PKEMRA) which sought to address the failures related to Hurricane Katrina response and recovery, specifically ensuring that FEMA would not fail in logistics. FEMA's mantra was to "go early and big" for every disaster. This led to over-extending FEMA with a myriad of small and medium size disasters.

These disasters, internal business practices, and an emergency management culture on the brink of collapse, caused Brock to ask some insightful questions and recognize the need to change the enterprise to include not only FEMA but also the profession of Emergency Management. His response to unprecedented disasters and an organization stretched beyond its capacity was to lead FEMA through a strategic planning process known as "Discovery Change Leadership" which entailed asking three questions: *Where are we? Where do we need to be? and How do we get there?* He invited input from leadership throughout FEMA through forums and digital processes, as well as external stakeholders (e.g., State emergency coordinators, National Emergency Management Association, International Association of Emergency Management, Nongovernmental Organizations, and the private sector). FEMA received thousands of comments, conducted a trend analysis, and distilled the input down to three goals on one page. This enabled Brock to build unity of effort and communicate it internally and externally to stakeholders. The resulting strategic plan enabled him to change the momentum and reestablish expectations. The remainder of this profile outlines how this strategic plan and its goals enabled Brock to lead FEMA through hundreds of disasters and ongoing crises despite stressors, organizational constraints, political challenges, and institutionalism.

The new mission statement of FEMA became: "helping people before, during, and after disasters." This simple, memorable, and visceral statement inspired a new era and purpose. Equally memorable, the mission seeks to achieve the vision of "a prepared and resilient nation."

The new strategy included three goals and realigned the agency's resources, personnel, and budget with this plan:

1) Build a culture of preparedness
2) Ready the nation for catastrophic disasters
3) Reduce the complexity of FEMA

A culture of preparedness begins with individuals. While FEMA had previously encouraged the public to stock up on 72 hours of essential supplies, there are many people that cannot afford to do this or fail to prioritize this as an essential element of personal preparedness. More people filed for Individual Assistance following disasters in 2017 than in the past decade. A growing lack of insurance has shifted disaster costs from individuals to the federal government. Nationwide estimates that about two-thirds of American homes are underinsured [2]. These problems are mounting, but Brock was also seeking the root cause. Leaders cannot focus merely on the problem (e.g., lack of preparedness or insurance) but need to address root causes in order to solve the problem. Brock has insightfully identified "asset poverty" as a major root cause. "Financial resiliency is desperately needed." Many people live on their income and accumulate debt that exceeds their assets. According to Prosperity Now (data from 2016), the asset poverty rate in the US was 24.1%; furthermore, 37% of households were liquid-asset poor, including 58% of black and Latino households. 47% of Americans had sub- or near-prime credit scores below 720, making credit inaccessible or disproportionately more expensive than those with credit scores above 720 [3]. Asset poverty is adversely affecting individual preparedness.

> **Asset poverty** is an economic and social condition that is more persistent and prevalent than income poverty. It is a household's inability to access wealth resources that are sufficient to provide for basic needs for a period of three months. Basic needs refer to the minimum standards for consumption and acceptable needs [4].

While a culture of preparedness begins with individuals, the whole community has an obligation to develop a culture of preparedness. This includes the public and private sectors. At the local level, emergency management recovery planning is lacking. There is too much focus on incident command because of lessons learned during Hurricane Katrina. Federal grants focus too much on National Incident Management System (NIMS) compliance. While NIMS and response are important, it is a small portion of the overall incident compared with the recovery effort, cost, and impact.

Key to success: Brock Long built and socialized the strategic plan in the field and with the media using the mantra, "This is our plan, not just mine." He encouraged the whole community of partners to adopt the goals to solve problems.

The Stafford Act, dynamics of responsibility, and ownership mentality have created incentives for the public and communities to fail in preparedness. Communities are rewarded for being uninsured and adopting substandard building codes. The Stafford Act has not established incentives for communities to do the right thing – i.e., implement real mitigation, strong building codes for residential and commercial structures, and insure public infrastructure. Additionally, the disaster declaration threshold is too low which shifts the cost and onus from the states to FEMA. This has led to historically excessive spending on recovery. For example, in 2017–2018, recovery spending exceeded the cumulative amount spent on recovery since the inception of FEMA in 1979. These factors challenge a culture of preparedness at the local and state levels of government.

In response to the culture in place within local and state government, Brock advocated for the Disaster Recovery Reform Act (DRRA). This effort took eight testimonies before Congress to emphasize the need for pre-disaster mitigation. Through the DRRA, Congress addressed several of the issued related to local and state preparedness cultures. Here are a few of the key highlights:

- FEMA provides assistance to state and local governments for the administration and implementation of building code and floodplain management as well as enforcement.
- Authorizes the National Public Infrastructure Pre-Disaster Mitigation fund, which is funded through the Disaster Relief Fund as a six percent set aside from estimated disaster grant expenditures. This allows for a greater investment in mitigation before a disaster. This new program is known as Building Resilient Infrastructure and Communities (BRIC).
- Requires FEMA to provide a regular report to Congress on the times and estimated amounts in which self-insurance amounts were insufficient to address flood damages.

The second goal of the strategic plan was to "ready the nation for catastrophic disasters." The biggest catastrophe faced by FEMA during Brock's term as administrator was Hurricane Maria's impact on Puerto Rico. Catastrophic planning by those in Puerto Rico was insufficient for the magnitude of this disaster. Several contributing factors exacerbated this disaster such as a neglected power grid, differing response philosophies in planning and response, and the inability to restore essential and critical infrastructure, such as the ports. In order to access Puerto Rico and provide federal support, the US Marines were deployed to reopen and operationalize ports of entry. Rather than integrate the private sector into preparedness and response, the territory looked to FEMA instead of accessing the 30-day supply of food and water which were accessible through grocery stores. Sadly, the grocery stores were not involved in catastrophic planning. As political and media pressure mounted over a perceived lack of food and water, the private sector resources needed to restore power and communications were displaced from the initial flights, and replaced with food and water supplies. This decision highlights how stress and external organizations such as the media and government officials can influence a crisis leader's decision-making. The problem was a decimated power grid and communications infrastructure, not lack of food and water. One lesson learned was the need to assess food supply and its associated duration, and then pre-deploy supplies before the incident when there is a warning time, as is the case with a hurricane. While there were many other lessons learned, one major outcome of this disaster was the development of Community Lifelines doctrine.

Community Lifelines answer the question, "What has to be working in a community, that if it's not working, then people die, or life routine is disrupted?" 85% of the nation's infrastructure is owned by the private sector, and not within the control of the government. Therefore, public–private partnerships need to address the needs of the private sector in restoring critical infrastructure. Additionally, the traditional Emergency Operations Center reporting process from each of the 15 Emergency Support Functions (ESF) creates information overload and does not prioritize information needed to make swift and impactful decisions as the number of major disasters increases. Each morning at 08:30, Brock would receive briefings from staff in ESF format. "This was like receiving an encyclopedia of information," he recounted, "too much information." He would speak with eight different governors from CA, TX, LA, FL, GA, SC, VI, and PR regarding the ongoing disasters. To streamline this process and focus on priority decisions and actions, he changed the reporting mechanism to the Lifelines construct and requested that the states provide information based on condition (i.e., green, yellow, red – see Chapter 7 on Lifelines for more information). This enabled him to prioritize actions based on Lifelines that were in red or yellow condition. Resource prioritization and decision-making focused on improving the condition of the Lifelines.

Through the implementation of seven interdependent Lifelines, reporting, decision-making, and prioritization became streamlined. Chapter 7 elaborates more on the Community Lifelines, application, and implementation at the local level. Each of these Lifelines requires coordination between the public and private sector in order to stabilize and restore the Lifeline. Instead of asking what the private sector can do for the government, the public sector should be working with the private sector to provide them with what they need to stabilize and restore the Lifeline for the community. To ensure the needed coordination across sectors, FEMA revised ESF#14 from "Long-term Community Recovery" to "Cross-sector business and infrastructure" to ensure unified and coordinated response and recovery. This fostered unity of effort between the public and private sector.

For example, the Food and Water Lifeline should operate under a vendor-managed concept in concert with major grocery store chains and bulk water suppliers and purveyors. The Housing Lifeline should strive to keep people where they are located rather than move them away from home and temporarily house them which is cost prohibitive and disruptive to their lives. A FEMA trailer costs between $200–300 thousand to setup, connect to utilities, maintain, and dispose, whereas, repairing a home during the aftermath of Hurricane Harvey averaged $60,000.

Addressing the Power and Fuel Lifeline often corrects problems across other Lifelines. During the COVID-19 pandemic, the medical supply chain imploded. Hospitals don't store excess equipment and supplies; they order them on a just-in-time basis. A lack of continuity of operations plans and redundant supply chains disabled this Lifeline. Local vendors depend on a few national vendors who were dependent on companies located in China and Malaysia for supplies. Once these supply chains were disrupted, the vendors could not provide the resources needed to the hospitals.

The third goal of the FEMA strategic plan was "reducing the complexity of FEMA." This required focusing on improving internal processes and serving the people of FEMA. Brock stated that, "good leaders are rooted in love and service to other people." Brock walked the facility and would ask people, "if you could change anything about FEMA what would it be?" His approach to servant leadership demonstrated that he cared and invited people to bring issues to the forefront that needed to be addressed. This started with small issues and then led to disclosure of bigger problems. He would take the time to listen and follow up with those that proposed new ideas, then set up a meeting with those responsible for consideration and implementation of the ideas in order to empower his leaders to take ownership of and show support for welcoming new ideas. This facilitated collaboration and a culture of improvement and problem-solving, which resulted in outing a leader who was sexually harassing members of the office, revitalizing Information Technology lifecycle planning, and hiring new leaders. Brock's leadership exemplifies the need to take care of the people supporting operations even in the middle of responding to disasters.

> "Good leaders attack the root cause of problem. For far too long leadership had a patchwork approach to solving problems."

Each chapter contains the "Keys to Resilience" based on the topics covered. While interviewing Brock Long, he shared his perspective on enhancing resilience.

Resilience is measured in dollars but since asset poverty is the leading root cause for lack of individual preparedness, we should consider the comprehensive credit score of a community as the community's resilience indicator. This is a lagging indicator for resilience, but a leading indicator for civil unrest, divorce, disasters, food deserts, and underinsurance. Some government leaders want to solve this problem with more police, but we need financial resiliency at the individual and household level. Focusing efforts on increasing one's credit score or home ownership is not addressing the root cause of the problem. The law does not incentivize people to do the right thing in preparedness – i.e., prevention and mitigation [5]. What is more expensive, billions of dollars to recover from a disaster or mitigation?

Our nation needs to address tough issues and change the narrative. For example, any house can flood. Imagine if home appraisers rewarded mitigation measures (e.g., hurricane shelters, elevated HVAC systems, etc.). The unfortunate narrative is that flood insurance is too expensive (i.e., $300–700/year). This is an issue of priorities, not affordability. Brock recognized that Congress was not going to fix the National Flood Insurance Program which is $16 billion in debt. Premiums are based on Congress' definition of affordability, rather than based on risk. During his term, FEMA was responsible for and unable to pay the interest on this debt. The program was insolvent. So, Brock took an innovative approach to solving this problem by engaging the reinsurance industry to backfill uninsured properties which could ultimately become a taxpayer's responsibility if disaster struck uninsured homes. FEMA paid $124 million for reinsurance in 2017, then Hurricane Harvey hit. The insurance policy paid out $1 billion in claims. Then, he continued to expand the program because the reinsurance industry was willing to take on the risk. FEMA should be focused on the uninsurable infrastructure that the US needs.

To change the narrative, Brock uses an analogy of a chair with four legs. The seat of the chair is the community which requires four legs for stability: (1) true culture of preparedness, (2) strong state and local government, (3) recognizing that 85% of the infrastructure is owned by the private sector, and (4) the firepower of federal government to be coordinated through states. All response to incidents is locally executed, state managed, and federally supported. Things can go wrong when one leg is missing. After Hurricane Katrina, there was only one leg: FEMA. The same was true with Hurricane Maria, there was only one leg: FEMA. During Hurricane Irma, a Category 4 Hurricane, there were more people registered for FEMA Individual Assistance, but all four legs were in place. During Hurricane Harvey, all four legs were firmly attached. Neighbor helped neighbor, demonstrating that there was a culture of preparedness in Texas.

Brock's two-year term with FEMA entailed hundreds of disasters, significant changes to the organization, and several crises from within. To say that this was stressful would be an understatement. Nonetheless, Brock faced the stress head-on knowing that he had to continue every day because his organization was counting on him to lead. That motivated him to suppress the stress of the situations and move forward. He spent time with his people, encouraged, supported, and congratulated them for wins; he had their backs. He took personal responsibility when things went wrong.

Brock recommended the following to prepare crisis leaders for the future:

- Surround yourself with good leaders and competent people whom you trust; this is biblical.
- Recognize that you don't have all the ideas.
- Trust, but verify information.
- Exercise a realistic span of control – i.e., five to seven direct reports.

"in abundance of counselors there is victory" Proverbs 24:6b ESV

Finally, to set conditions for success during a crisis,

- Prepare in advance by informing plans with Lifelines.
- Establish pre-incident contracts needed to provide staff augmentation.
- Obtain technical expertise needed.
- Procure or gain channels of access to equipment.
- Prior to requesting state resource, exhaust local and mutual aid resources.
- Establish vendor-managed concepts to reduce dependency on state and FEMA resources.

10.3 Lieutenant General (Ret) H.R. McMaster: Counterinsurgency against Al Qaeda in Iraq

The section is based on an interview that I conducted with General H.R. McMaster on May 13, 2021.

Lieutenant General (retired) H.R. McMaster is a Senior Fellow at the Hoover Institution at Stanford University. He served as National Security Advisor to President Trump from 2017 to 2018. His military service included 34 years in the US Army and he is the author of *Battlegrounds: The Fight to Defend the Free World* [6] and *Dereliction of Duty* [7].

In 2004, US forces drove terrorists out of Tal Afar, Iraq, but failed to leave sufficient troops in place to hold the city. Al Qaeda in Iraq (AQI) retook the city and incited a civil war between Tal

Afar's biggest ethnic group, the Turkomen, by turning Sunni against Shiites. They established brutal control over the city with extreme acts of violence and torture. They closed schools, markets, and executed police by death squads for retribution. The people of Tal Afar suffered the brutality and violence of AQI. Tal Afar became the AQI center for training. They had courses on kidnapping, murder, and surveillance. AQI was organized and sophisticated.

Col H.R. McMaster, Commander of the 3rd Cavalry Regiment was tasked with the mission of leading US and Iraqi forces to prepare major offensive operation and restore security to the city under Operations Restoring Rights. This operation was conducted from September 1–18, 2005.

The stakes were life or death. Col McMaster had witnessed the brutality, bombings, suicide bombings, and heinous scenes. Loss of soldiers and lack of sleep added to the stress of combat. He faced dangerous situations against an allusive enemy armed with roadside bombs and snipers.

But Col McMaster remained focused on his responsibilities to his soldiers, to the people of Tal Afar and achieving a favorable outcome. The sacrifice was worthy of the risk and this helped insulate him from the stress of combat. He described it as "stoicism" and an ability to "suppress emotions that are not helpful."

Throughout the operation, he maintained situational awareness by establishing security in neighborhoods instead of the prevailing tactics. He established relationships with those that suffered and gathered Human Intelligence (HUMINT) on the enemy. This helped enhance his visibility on enemy's organization and leadership. Then he capitalized on a key source that his forces captured, an abused adolescent who had been dehumanized and passed around four battalions throughout the AQI organization. He knew the organization well. This young man had been sent out on an assassination mission but was captured by US forces and treated well. During his debrief, he broke down and provided comprehensive intelligence on the AQI organization. Col McMaster supplemented his intel with knowledge of history, culture, and lessons learned from the failed Operation Black Typhoon, an attempt to defeat AQI in September 2004. Situational awareness continued to increase as Col McMaster launched major operations in advance of the main offensive; he raided 20–30 targets simultaneously which led to the capture of individuals of interest and new intel.

Good intel enhanced situational awareness which improved decision quality. Col McMaster made several important decisions. The first was to evacuate civilians; this allowed fire power and avoidance of innocents. He met with local leaders to build trust and a relationship so that they could focus on the common enemy. As they evacuated civilians, they screened evacuees and detained suspects.

Second, he cordoned off towns in areas where terrorists would flee. By conducting humanitarian missions within the city and surrounding towns, they gained access, focused on a common need, and photographed every adult male so that they could pursue any would-be terrorists that escaped.

Third, he built a 10-foot-high berm around city, established three check points in and out of the city, and placed informants at check points. These measures restricted reinforcements.

Finally, he conducted reconnaissance around Tal Afar, border surveillance, and interdiction operations to stem the flow of reinforcement from Syria.

These decisions set the conditions of the battle and led to enhanced situational awareness. But there were still uncertainties. Col McMaster remarked, "War is profoundly uncertain due to countermeasures of deception and influences." Commanders must be comfortable with uncertainty, triangulate to find ground truth, and verify information. Commanders cannot be paralyzed by waiting for more information to increase certainty. The riskiest course of action is to do nothing as the that allows the enemy to take the advantage. Col McMaster believes that German author, von Clausewitz said it best in *Principles of War* [8], "We must boldly advance into the shadows of uncertainty." When navigating decisions, Col McMaster advised, "Look for opportunities, determine who else needs to know, seize, and exploit initiatives. Some are paralyzed by dangers and blinded to opportunities."

Throughout the operation, Col McMaster had to win the influence from within the US government, tribal leaders in Iraq, the people of Tal Afar, and the international audience. To win allegiance with the people of Tal Afar, he emphasized their common humanity and traced grievances back to the enemy. He learned that the people were frustrated by Shiite-dominated government that did not represent their interests. They were upset with the lack of utilities such as a reliable water supply and power. He learned that the enemy would kill the water-truck driver and every time power was restored, they would take out a transformer. AQI would prey on the Sunni Turkomen population. Whenever there was a recruiting drive for new Sunni police, they were afraid to join. Col McMaster became a mediator for the people by removing Iraqi leaders that were an impediment to security (or in some cases, getting them promoted to Bagdad, or convincing AQI sympathizers that they were better off in Turkey). He was able to win the trust of the people by convincing them that the US was working for them and ultimately reforming and refurbishing the police force.

Col McMaster also had to wage a counter-information war against Turkomen who were connected with AQI and were sending misinformation to Turkey that US was victimizing the Turkomen. He worked with the US Embassy in Ankara and invited Turkish forces to shadow US forces in order to provide first-hand accounts of their activities. He worked to attract other Turkomen forces to join the fight against AQI. He leveraged his network and relationship with General Petraeus to reallocate needed resources such as armor protection to local Iraqi forces. He flew to the Iraqi parliament to share about the work that the US was doing. He established a relationship with Robin Brims, the British Deputy Commander and liaison to the Iraqi Parliament who also became the US liaison to the Iraqi Government. Ultimately, he learned more than the people who had lived in the Nineveh Province so that he could establish his credibility and counter the misinformation from the enemy. To address the international audience, he invited journalists like Michael Ware, but unfortunately *Time* did not post the photos that were taken, perhaps they painted a heroic portrait of US soldiers.

Col McMaster had to convince the US headquarters that Tal Afar would not become another Fallujah. Since his boss was not convinced, he pursued the next commander in his chain of command to ensure the right information was getting to the right decision maker.

As a leader, General McMaster has found the following attributes and skills to be essential in successfully navigating a crisis:

- Listen and learn about complex problem-sets [9] empathetically. Frame complex problems before rushing to action. In *Battlegrounds*, General McMaster highlights the importance of strategic empathy, a view from the perspective of others.
- Exert influence, understand your limitation, and cooperate with others that have different perspectives and assets.
- Gain an interdisciplinary perspective and apply design thinking [10]. Embrace diversity.
- Establish clear goals, objectives, and take advantage of opportunities.
- Exercise effective communication skills, rather than hierarchical reporting.

General McMaster recommends that leaders preparing for future crises should read history, learn from the experience of others, and learn to ask the right questions. They should build strategic competencies and gain an understanding of contemporary challenges through the context of history.

10.4   Frank Patterson: Incident Commander during the West Fertilizer Company Incident Response

This section is based on an interview that I conducted with Frank Patterson on April 2, 2021.

On April 17, 2013, a fire and subsequent explosion at the West Fertilizer Company in West, TX, killed 12 responders and 3 members of the public. This incident of national significance injured 262 people, damaged 150 offsite buildings, and resulted in the loss of $230 M. The explosion of approximately 50 tons of fertilizer-grade ammonium nitrate occurred only 20 minutes after the first signs of a fire were reported to the 911 dispatch center. Initial response lasted throughout day one until all 262 patients were transported to the hospital (about 8–10 hours). Other response actions lasted a couple days. Recovery lasted a few years.

Frank Patterson, emergency manager for Waco, TX, and McLennan County, TX, was requested by Tom Mark, the West EMS chief to serve as the incident commander until the mayor officially appointed him as the EM coordinator on day two (until official appointment, he functioned in a mutual aid capacity). When Frank arrived on the scene, many personnel were already responding and his first task was to size-up the situation, organize responders, create strategic objectives, and prioritize resources to accomplish the objectives.

Frank managed the incident by driving objectives into resolution. The various impacts of the incident (i.e., casualties, damaged buildings, hazardous materials, etc.) drove the need for branches within the ICS. He organized, built the team, and assigned responsibilities according to respective roles.

One key to managing the crisis was recognizing that anything done to stabilize the response affects recovery; so, in practice, the response and recovery phases overlap considerably. He remained engaged in response and recovery for several months until July 2013 when the federal investigation concluded and was transferred to the state fire marshal's office.

There were several stressful aspects encountered throughout response and recovery. Initially, a lack of situational awareness (i.e., not knowing who was there, what activities were taking place, accountability on casualties, and their status) caused some stress. The magnitude of incident was very large and overwhelming. So, considering how it would be resolved and with what resources contributed to the overall stress of the incident. Frank recalls feeling like there was an unending line of casualties requiring transport to a hospital as the number of casualties was unknown and changing throughout the incident. Since "life-saving" was the primary strategic objective, this consumed most of the resources and mental energy. Fortunately, search and rescue operations were not uncovering additional casualties in impacted homes. But when the final patient was transported, the crisis shifted to other priorities related to response and recovery actions.

Another major stressor on incident command was the spontaneous arrival of volunteer, state, and federal resources that had not been requested. This quickly became a management issue and stressor for the IC. The incident quickly garnered national attention and came into the political spotlight. On the roadway behind the staging area, elected officials gathered looking down on the scene and those managing the incident. Eventually federal workers arrived from multiple agencies; this diverted attention from incident management.

Throughout the incident, Frank did not experience noticeable physiological or psychological signs or symptoms associated with stress. As an experienced firefighter and emergency manager, he knew what to do, relied on prior planning, education, experience, and training to execute the mission. Reflecting on his role, he stated, "Leaders need to be comfortable in a role and not let external influences affect response." It was not until September 1st (four and a half months later) that he really reflected on the incident and his role in the response.

While there were stressors during the incident, most of the stress did not come until the investigations after the incident.

Frank noted three keys to managing stress during the incident: (1) he assessed and reassessed the size and scope of incident; (2) he was able to compartmentalize elements of the incident response as well as his personal life, and (3) he was able to control the ops tempo by slowing down the incident in his mind to think through the process methodically. He also recognized that he needed to be decisive and avoid paralysis by analysis or trying to make the perfect decision as this threatens a leader's ability to make good quality decisions. He would make the best decision based on best available information at the time.

Since the incident command post was in field conditions and far removed from scene, he was more reliant on situational awareness reports rather than IT-based tools. His decision-making process followed four basic steps: pause the response, receive reports, prioritize, and decide. By pausing the response, he was able to receive and share information consistently across operations. This only took a few minutes for everyone to sync up across functions. Then, they would adjust as needed. Briefings helped keep people informed to prevent freelancing. Not only did the primary responder participate in the briefings but he also used this opportunity to provide updates to everyone engaged in the response, to include volunteers. When he was informed that several vehicles had flat tires, he was able to rely on his logistics section chief to address it by obtaining the needed resources. "Leadership involves trusting people and letting them do their job."

The briefings also revealed some coordination and alignment issues with external parties. In some cases, the briefs realigned these groups with the incident objectives, but in other cases, Frank had to make the decision to demobilize some teams and personnel. For example, on the second day, he needed to demobilize an IMT because they were not integrating into the system and moving incident forward. At the end of the first day, he demobilized volunteers as they were not needed; however, many of the volunteers refused to leave and stayed until the next day in case they were needed. He did not make this an issue but rather focused on accomplishing incident objectives.

The primary tool available to each section/branch of the ICS was radio communications between the branches and section chiefs. The section chiefs reported directly to IC with verbal situation reports since they were in the middle of a field. As the response developed, he requested a command trailer for planning and logistics sections.

There were several key decisions that Frank had to make throughout the crisis. These included:

- Decision on which hospital to send casualties based on impending severe weather. As a strong frontal system approached during the first evening, tornado watches and severe thunderstorm warning threatened the ability to evacuate casualties by air, so Frank made the decision to deviate from local plans and send patients by air to the farthest hospital so that if severe weather arrived, they could focus on transporting patients to the nearest hospital by ground transportation.
- As media helicopters flew over the scene, they were impeding response operations due to airspace conflicts. Therefore, Frank requested a "No fly zone" over the incident in coordination with the FAA.
- While responders established triage within a local football stadium at an early stage because of its capacity, lighting, proximity to the incident, upwind location, and accessibility, the winds shifted as the frontal system approached the scene. This placed the football stadium downwind of the incident. So, Frank made the decision to relocate the triage staging area.
- After the first person died in triage, Frank ordered a refrigerated truck to handle the deceased. This was a very difficult decision as it acknowledged the realization that this could become a mass fatality event. This was a mental stressor and a challenging turning point in response operations, but a necessary decision.

- At an early stage, Frank established incident objectives. The highest priority was life-saving. As they transported the last known patient, this left a memorable impression of relief that they had done all that they could to save lives. Now the focus shifted to firefighting, Hazmat response and search and rescue operations. This was a huge turning point from crisis to de-escalating response and recovery operations.
- The command post had setup in the field but by 2 am, Frank made the decision to move into a church building to provide adequate shelter.

While many leaders become distracted or influenced by the media or what others are saying about the leader's decisions, Frank asserted that he would not let the media or external parties influence his decision-making. Despite even the advice from his wife to "get a handle on the media," Frank focused on doing the right thing at the right time. He did not attend or hold media briefings but deferred to elected officials in handling of the media. But even the elected officials were busy with engaging in the incident. Nonetheless, some media outlets proliferated false information. While some of this could have been prevented with earlier engagement with the media, these reports did not affect response operations and priorities but rather created false narratives which needed to be addressed during investigations.

While Frank was able to obtain good, reliable information on the status of the main response processes, there were instances when he would need to process available information, make a decision, and adjust as needed. For example, the decision to establish a triage staging area in the football stadium was based on the best available information and was a good decision at the time. However, when winds shifted, he adjusted and made another decision to move the triage staging area. This decision continued to move the incident forward. By staying focused on strategic priorities and incident objectives, he was able to navigate the incident to its conclusion.

The most positive factor that affected Frank's ability to lead through the crisis was the trust and familiarity with other responders. They knew each other, understood each other's expectations and capabilities, trained together, and exercised together. They were able to leverage each other's strengths. Because of this trust and mutual understanding that existed prior to the incident, Frank was able to confidently delegate without micromanaging. As Frank stated, "many leaders can micromanage and still be unsuccessful." He just needed to know outcomes from operations and decisions made.

A major detraction from his ability to lead and manage the incident was caused by unsolicited state and federal resources deploying to the scene and failing to follow the national incident management system – locally executed, state managed, and federally supported. Several of these resources did not come into alignment with the ICS and incident action plan, but instead challenged these by initiating their own priorities. This made leading a unified response more challenging. Some of these external resources began meeting individually, neglecting meeting with the rest of the responders, so Frank had to get them realigned. In some cases, he had to demobilize teams and order them to leave the incident. In other cases, these teams recognized their role in the response and aligned with and supported the incident objectives.

The biggest challenge to leading the response was "personalities, egos, and politics." Frank's advice to aspiring crisis leaders is to "get thick skin, and don't take it personal." Realizing that everyone thinks their job is most important helps leaders calibrate and customize their communications such as incident priorities among team members and achieve alignment. A leader needs to facilitate conversations across teams and lead the group in its entirety towards common goals. Sometimes, this requires a leader who can shake things through healthy confrontation to highlight the misalignment that needs to be addressed.

While everyone has the ability to lead at some level, not everyone is able to become a good leader in a crisis. Some leaders simply will not get to this level. A title does not make you a good leader. There is a difference between a crisis leader and a positional leader.

To be a good leader, one needs a high degree of self-awareness of whether they have the traits required, skill set, and ability to lead in a crisis. A good leader understands their strengths and weaknesses, inspires others to follow them, and listens well, even to the most junior person. A great leader is a good follower, knows and appreciates their profession, has experience leading in a crisis, is decisive, and has gained sufficient experience to understand what their colleagues within the operation are explaining and accomplishing. Leaders create an environment to foster leadership development.

Frank highlighted that in addition to leadership skills, leaders need to possess certain character traits which establish the foundation for success. These include integrity, respect for those around you, appreciation of other skill sets, ability to instill confidence in those you lead, and boldness.

A crisis leader requires all of the skills and traits mentioned above, but also must have the experience, education, and training. While a crisis leader will inevitably be nervous going into their first incident, they need to hone their skills through smaller incidents. Frank highlighted his method of slowing down the pace of the incident to control the ops tempo and compartmentalizing the incident into smaller pieces to make it manageable. "Don't worry about the whole, just start eating the elephant. Don't get overwhelmed by the magnitude of the incident, it can cripple some people. And remember, the leader did not cause the incident, but don't make it worse." People are looking to leaders to make decisions; this is the primary expectation. Do not get paralyzed by analyzing all of the consequences or attempting to make the perfect decision. The wrong decision is "not making a decision. Do something."

From the outside looking in, this event ran smoothly. On the inside, there were battles that were dealt with internally. A good leader can differentiate the two and keep private, private; a good leader will not publicize disagreements following the adage, "Praise in public, criticize in private." People will follow you if they trust you. Respect one another. Listen. Position does not give authority; other people give it to you.

Aspiring crisis leaders should focus on gaining experience. Book knowledge does not deliver understanding until you have experienced a crisis. Without experience, leaders do not understand the challenges and will be ill-equipped to handle a major crisis. This requires existing leaders to invest in developing leaders by allowing them opportunities to gain the experience.

The keys to setting conditions for success start before a crisis with establishing integrity, professionalism, trust, understanding leadership styles (and which will not work in a crisis), learning to be decisive, and understanding various responder roles.

During a crisis, avoid mission creep and keep everyone in their respective lane. Be confident in decision-making. Recognize and learn from your own mistakes. Demonstrate humility so that lessons can be learned, not lost. History repeats itself because of lessons lost. Finally, recognize better ideas from others. A leader with an ego prevents the best ideas from creating the best outcomes.

10.5 Derrick Vick, President of Freedom Industries: Leading through a Financial Crisis and a Return to Prosperity

This section is based on an interview that I conducted with Derrick Vick on May 12, 2021.

Derrick Vick is the President of Freedom Industries, based in Rocky Mount, NC. Previously he served as the Chief Financial Officer for 11 years. During this, time he worked closely with Doug Ezzell, the company's CEO. In January 2021, Derrick became president as part of a leadership transition. The original founder retired shortly thereafter in June 2021.

Freedom was formed in 2004 and is a 100% employee stock-ownership plan (ESOP) company. Freedom operates in the Mechanical, Electrical, and Plumbing (MEP) sector with a specialization in heavy industry focused on food, pharmaceutical, auto, and heavy manufacturing. This diversification provides an element of recession-proofing which has enabled growth during normal recessions.

In 2014, after its conversion to an ESOP, Freedom experienced record years in revenue growth. They had dialed in on the Electrical segment of their market and this created solid business opportunities. To further capitalize on this strength and diversify within their market, they entered the Mechanical market as there were many opportunities based on the design of industrial plants. Because the mechanical market looked promising, Freedom began building a backlog of work and expanding their team.

As the team continued to grow, so did the revenue. A $2–3 million backlog became $20 million. The financial analysis looked solid and made good sense, so Freedom hired more skilled tradesmen to continue expansion into the mechanical market.

Then everything began to change. Approximately half-way through their first major Mechanical commercial project in March 2018, overhead began to balloon into millions of dollars. A project lifecycle is typically about 12–18 months. So, this became very concerning given the remainder of the project as well as the large backlog of new work that they had booked. Gross profit started fading. Freedom began to incur project delays and they were losing over $1 million on a $25 million revenue year. Then projects started falling apart and the excuses and blame shifting began. Freedom had allowed the backlog of work to grow too quickly before ensuring the company had the capabilities to conduct the operations.

Unlike the Electrical industrial market which was based on quality and relationships, the low price, slim margin work in the Mechanical commercial market was unforgiving and exposed Freedom to extensive financial exposure and risk. As excuses accumulated, the Board weighed in on the matter and identified that the new direction was no longer tenable. However, the leadership was faced with their first major decision and dilemma: Freedom had staffed a new division with knowledgeable leadership who were intimately involved in the projects that they had captured so they needed to retain this team to continue managing the projects but did not want to book more sales within the commercial mechanical market. The Freedom leadership team reached a decision point: abandon this new market so that they could refocus their business on the Electrical industrial market. However, their culture was upended during this period of growth. Workers were upset about the change in mission, hiring new employees at a premium, and new leadership; so, the leadership team set a course to "return to prosperity."

Ultimately, the Freedom leadership team made the decision to refocus their energy on the Electrical industrial market segment. However, this left many members of the company feeling upset because the culture was upside-down. Many felt as if the mission of Freedom had suddenly and drastically changed. It was crucial that the leadership team set a course to return to prosperity.

Fast-forward to June 2019. Freedom had other problems brewing. They were entering a cash crisis; their line of credit was fully utilized; cash and profits were dwindling. They were over 100 days late in paying vendors. Superintendents were cut off by suppliers and this inhibited work, creating a snowball effect (i.e., the effects of one problem had consequences that led to another problem creating compounding effects). This negative information began circulating among the employees.

One crisis led to the next crisis. Freedom's leadership team developed a strategy on expense reductions and leadership changes because they needed to address the floundering division as they attempted to close out their mechanical backlog of projects. Cash was tight and it was a matter of time before they would expend their remaining cash and face bankruptcy. At this point the death of the company was academic. Fortunately, the leadership team was not going to give up easily.

The leadership team, led by Doug and Derrick, outlined Freedom's "Return to Prosperity" plan which included selling his accounts receivables, transitioning business development back to

Freedom's core competency in Electrical work, increasing margins, and hyper attention to cash-flow management. He established a new strategy and communicated the situation, the errors, and the new strategy transparently with his team members.

Freedom raised new capital, opened a new line of credit, and grew their Electrical division backlog to offset losses from their Mechanical division. They searched for new financing but the time to access lending is not in a time of crisis. Their primary bank brought in corporate to take over their loan accounts. Freedom presented their business case and recovery plan to the bank, but the bank gave Freedom 30 days to eliminate their debt or the bank would call their lines and collateral (i.e., seize their assets if they can't payback their loans). Derrick contacted 11 different banks to no avail. As a last resort, he devised a triparty plan with two banks and his company to sell his accounts receivables (i.e., the bank takes ownership of payments due to the company in return for a fraction of the amount due; it is not cheap!). This triparty agreement gave Freedom six months to survive and get back on feet with positive cash flow.

But how would he turn the tide in six months? And what indicators could he use to manage this crisis and lead to a successful outcome? Each crisis is different and requires a unique set of indicators to inform leaders and decision makers about the situation. In this case, cash-flow projection was the leading indicator needed to create situational awareness and inform decision-making. Whereas the income statement was a lagging indicator of profitability. Derrick maintained and monitored in real time his cash-flow projection for 12 weeks into the future. To account for uncertainty and late payments he utilized historical, empirical data on gross profit and accounts receivables to adjust his projections. This indicator clearly showed the problem, but he needed to act in order to reverse the gloomy cash-flow projection to bring in more cash than he paid out in expenses. This was the pivotal point in this crisis. Hence, he decided to change his priorities so he could increase attention to improving cash flow. This required deferring or eliminating good tasks such as reporting in lieu of the number one priority – improving cash flow. He also needed to establish a company-wide strategy with multiple objectives to stabilize the sinking ship and return to afloat status.

He took a proactive approach to managing his accounts payable (i.e., bill paying) and receivables (i.e., collecting payments). He began calling all of his customers who owed money and offered them discounts for early payments; many were receptive. On the other hand, vendors were calling him demanding payments that were overdue. He explained the situation and negotiated lower payments and extensions. Unfortunately, the math still did not work out. Freedom was not generating enough cash flow to remain solvent. Nonetheless, Derrick never gave up. He continued to plow headfirst into this storm following the "buffalo culture."

Derrick Vick is a member of the C12 Group, an international network of peer advisory groups designed to help Christian business-owners "make better decisions, avoid costly mistakes, create solid plans for growth, and thrive in their calling to create impact beyond the bottom line." The C12 Group espouses the buffalo culture mentality of facing directly into uncertainty, risk, and instability. When the storm approaches, buffalo move as a herd into the storm instead of getting caught up running away from the storm. Moving into the storm results in less time spent in the storm, whereas cattle try to outrun the storm. Instead, the storm catches the cattle who spend more time in the storm, leaving them exhausted for predators. Buffalo move into the storm with purpose, direction, and mental toughness as a herd thereby reducing their risk of peril.

Learn more about buffalo culture at buffaloculture.com

Learn more about the C12 Group at www.c12group.com

The extreme stress of this crisis weighed heavy on Derrick. The math simply did not work out; they were heading for bankruptcy. He was losing sleep and was convinced that they would not survive. He had already engaged the bankruptcy attorneys. Meanwhile, he did his best to keep people engaged and the organization running. He carefully weighed how much of the foregone conclusion to broadcast to employees while he fought for a different outcome. During this stressful time, Derrick was never cut off by his suppliers as he continued to work down his payables and slightly change the momentum so that he could stay ahead of the bankruptcy.

When asked how he managed the stress and how it affected his decision-making, Derrick shared, "With lots of prayer." Derrick even organized and offered a time of prayer and fasting during lunchtime for his employees. He shared the challenges and issues facing the company and they prayed. His natural inclination was to delay difficult decisions, but an external Board of Directors was helpful in assessing the situation objectively and establishing decision points written into a plan. This removed much of the emotional toll and stress associated with difficult decisions and ensured timely decision-making according to a set schedule. They also presented this plan to the bank. He recounted that this external accountability associated with a written plan and time-bound decision points eliminated much of the emotional stress. Finally, the leadership team vowed to compartmentalize the stress associated with their work so that they would not bring stressors home to their families. He shared that this was difficult and required a high degree of self-aware-ness, but necessary to maintain a good life balance and home life. So, at the end of each day, they agreed to turn off business until they returned the next day.

From June through December 2019, Freedom implemented their "Return to Prosperity" plan steadily closing out contracts and growing their Electrical segment work. Then in December 2019, Freedom landed their largest project ever with their best client. The backlog increased and cash flow improved. Freedom fully recovered from this crisis in June 2020, almost three years after it began and two years after they realized that they were in a crisis.

This crisis included several external stakeholders such as the bank, Board, and suppliers. This created accountability and enhanced objective decision-making. Their Board offered sound advice that they needed to hear. It was hard but helpful. For example, the Board informed him that overhead costs were not enough and that their plans were missing essential items which needed to be addressed. Derrick also relied heavily on advice from his C12 peer group.

Derrick found that having a plan was essential to managing this crisis. While a plan is not permanent, it did establish objective criteria, decision points, and schedules for key decisions. He recommends keeping it simple and communicating it throughout the organization. He confessed that it was hard to cast vision during the crisis, but the plan helped him do that. "It was not a time to speak from the heart, we needed a methodical, precise plan on what the organization needed to hear," recalled Derrick. He established key talking points for the executive team to ensure they were communicating a unified message. Consistent communication was an essential element required from leaders. This demonstrates the necessity to have a plan and communicate the plan consistently in all situations.

Crisis leaders need to have a good pulse on leading indicators so they can project into the future and sense a pending problem. Rather than wasting efforts on solving small parts of the problem (e.g., cutting overheard), leaders need to focus their efforts on tasks that will have the right impact and effect on the situation.

"Facts are your friends," Derrick advised even though the facts said they were going bankrupt. Leaders need to take emotions out of the equation. Some leaders are more people-oriented and less analytical; this could be dangerous for the leadership team. Derrick believes they got into this crisis because they were too optimistic and did not see the facts. To avoid blind spots, leaders need wise outside counsel with a board, business association, and peer group comprised of people in similar roles (e.g., CFO, CEO, COO, etc.). Finally, leaders must "maintain confidence and joy in a time of crisis."

10.6   Chad Hawkins: COVID-19 Response Incident Management Team Incident Commander

This section is based on an interview that I conducted with Chad Hawkins on April 6, 2021.

Chad Hawkins serves as an Assistant Chief Deputy State Fire Marshal in the Oregon Office of State Fire Marshal. In late February 2020, as the COVID-19 pandemic began to exceed the capacity of the Marion County Incident Management System, the Marion County Emergency Management office requested an Incident Management Team (IMT) to assist with the enhancement of their current Incident Command System (ICS) structure. Initial objectives were to support the implementation of a rapidly expanding ICS organizational structure, develop processes related to the COVID-19 response, and develop an Incident Action Plan (IAP) needed to respond to the growing crisis and the complexities presented. Chad and his short IMT (resources were scarce at this time and a full IMT was not available) were deployed to Marion County and he served as an Incident Commander (IC) in a unified command structure with the Public Health Division Director until the end of March 2020. Early recognition of the growing crisis enabled Marion County to request an IMT and ensured they had the ICS resources in place needed to manage the incident. While the health department and emergency management office were familiar with previous outbreaks such as tuberculosis, small-scale public health crises, and all-hazards incidents, COVID-19 was becoming an atypical response and it was recognized that this would be a long-term complex incident.

During this deployment with the IMT, Chief Hawkins led approximately 75 staff members and leveraged his prior experience as an incident commander and serving in other ICS roles during wildfires, mass casualty incident (MCI), Hazmat incidents, and other disasters across the western coast of the United States. His experience and familiarity with the ICS enabled him to come alongside the public health department, establish the ICS organizational structure, develop a robust IAP and response options, and eventually transition control back to the locality.

At the early stage of the COVID-19 crisis, there was a lack of certainty regarding transmissibility and health effects. Chief Hawkins was presented with several key decision points and challenges during this deployment. An early issue at the beginning stage of the pandemic was recognizing that first responders still had to respond to routine emergencies which could result in exposure to public safety workers and potentially contaminate critical infrastructure and assets such as hospitals, ambulances, and police vehicles to name just a few. In order to protect essential public safety workers, they needed to source and allocate PPE to responders at potential risk of exposure to infected members of the public. So, where would they find PPE? Who would receive it? How would they distribute it and when? After all, PPE was a limited resource.

Chief Hawkins, Marion County personnel, and his IMT developed a methodology to address these questions. However, they hit their first hurdle. The national stockpile contained N95s but these required fit testing specific to the brand, and the N95 respirator brands varied throughout the stockpile. This created an interoperability challenge. Second, his team needed to determine the appropriate level of PPE based on exposure risk. As a Hazmat-trained firefighter, Chad was no stranger to selecting PPE based on the protocols outlined by NFPA and OSHA. However, these protocols are based on what is known about identified hazards, their quantity or concentration, and a body of toxicological knowledge. SARS-COV-2 was a novel virus devoid of any exposure science. This was a global pandemic with many uncertainties and variables. Unlike most incident responses, this was an invisible hazard. Nonetheless, Chad applied his knowledge and experience of unknown hazard situations. IMT members, local Public Safety Answering Points (PSAPs) and the state Emergency Communication Center refined screening protocols developed by Oregon Health Authority for both medical and nonmedical calls to assess the risk and make recommendations to responders. The largest stressor for him was knowing that first responders were transporting patients and responding to calls that had the potential of contaminating critical resources and infecting personnel and the public.

Having served in the field during many responses to large incidents, Chad understood the importance of employing the right resources to accomplish a task. He tenaciously pursued allocation of resources for responders to do their jobs, but in some circumstances, he was unable to provide the resource requests due to the limited supply. This affected him psychologically as he had never been in a position where he was unable to provide the required PPE to responders. It's important to note that this was a potential life-and-death issue for firefighters, and he recognized the same potential implications for other frontline workers. This furthered the need to prioritize limited resources, evaluate calls into the emergency communications center, and PPE distribution.

The first case of COVID-19 in Oregon was on February 23rd. By the time Chad deployed at the end of February with his IMT, there were hundreds of cases in Oregon. While the health department had a plan in place and the department operations center was established with personnel assigned to various positions, it became clear that the scale and scope of this incident required an extensive ICS structure beyond the capacity and previous experience of the County. Until COVID-19, the Health Department had typically staffed an ESF#8 position in the EOC during incident but did not need to implement the ICS to the extent of filling key positions. While they have highly trained public-health experts, many key personnel lacked the training and competencies necessary to perform their ICS roles and deliver products such as a robust Incident Action Plan. Chad quickly needed to assign the right people to the right positions so that he could assess the situation, obtain useful and timely information, and coordinate an approach to develop an Incident Action Plan. So, his IMT assumed the Planning Section and Operations Section Chiefs positions while simultaneously searching for qualified and competent personnel to transition into these roles when the IMT demobilized. His team provided direction, structure, and stability to the ICS establishing incident objectives and engaging subject matter experts as technical specialists in the ICS. However, capacity was limited. They did not have enough people to assign to roles. Additionally, some people were directly impacted by the crisis, becoming ill or needing to care for infected family members. They needed trained and available personnel in a resource-constrained environment. On-the-job training was an essential element to training personnel and volunteers. He assigned personnel from other disciplines as Unit Leaders and deputy positions within the ICS.

But the greatest challenge was getting ahead of the incident response in order to stabilize the incident and the incident management system. By working with the health department leadership to develop incident objectives, he was able to establish the task, purpose, and end state to create order and direction.

The stress from these complex issues manifested itself as a physiological effect. Chad recognized several digestive system effects – e.g., not eating, pit in his stomach, and upset stomach. He became tired due to the endless hours of managing the incident. When he recognized these stresses manifesting, he fell back on healthy coping mechanisms as much as possible. The nature of COVID-19 not only impacted his professional life, but also his personal life. The things we look forward to, like getting together with friends and family were suddenly out of the question. Because of the lack of high-quality information, changing nature of the situation, and many uncertainties, he recognized that this incident could lead to planning paralysis. Occasionally, he had to stop and hold a tactical pause, reset, engage with the technical experts, ask questions, and then reapply his knowledge of ICS and leadership to the changing public health crisis and then reengage.

Initially, there was a low level of situational awareness when he arrived. However, Chad was able to improve situational awareness by assigning the right people into the right positions to obtain the needed information. He leveraged external resources such as the State ECC's Common Operating Picture platform to increase their visibility on state-wide resources and requests. While during most incidents, communication issues are cited as a frequent problem or weakness, Chad noted that communication was much better on this incident than others. The whole of government approach and utilization of common operating picture platforms ensured dissemination of daily messages and briefs to enhance information sharing. Multi-jurisdiction information sharing ensured that localities were aware of resource availability and shortfalls. Public information was also well-coordinated and this helped with situational awareness. While overall communications were good, they became rich and redundant at times. Social media served as main conduit for information sharing since most people were at home quarantined. This highlighted the need for good social media presence and emphasis on digital information.

In the absence of completed information, Chad hypothesized projections into the future based on current data and scientist input on future projections. He recognized that not making a decision would be adverse, so decisions were needed to move the response toward positive outcomes. He would make decisions based on what information he had, not on what he did not have. There were several additional issues that hindered a truly effective and efficient response such as numerous conspiracy theories and political biases. First responders are not used to dealing with that level of political optics seen throughout this incident so this became an additional layer of complexity. Competing interests among jurisdictions caused tensions in decision-making and prioritization of resources. To address this, the County established a policy group early on and Chad supported the County by setting the leader's intent and objectives to facilitate engagement with political leaders.

To improve situational awareness, he would occasionally hold tactical pauses so that the response team could reassess the leader's intent, progress toward established strategic objectives, and evaluate the effectiveness of the associated tactics. These included working with liaisons with community groups to reach underserved populations. They scheduled public information releases twice daily, received routine updates on contact tracing operations, and established a PPE distribution methodology. Working with a team of epidemiologist subject matter experts (filling the Technical Specialists role within ICS), they established metrics which were reported daily before the tactics meeting. These metrics included number of cases, hospitalizations, deaths, and trends. These technical specialists also served as a conduit to the State and CDC; they reported directly to the ICs and were not embedded in functional groups which helped ensure their credibility and independent perspective.

In addition to establishing a PPE distribution methodology and establishing incident objectives, Chad had several other key decisions to move the response eventually into the recovery phase. Early in the response, Chad was able to project future requirements and begin to address them at an early stage to get ahead of the response. These included establishing and increasing the capacity of contact tracing and planning for mobile testing and mass vaccination in the spring of 2020. To ensure this work continued, he assigned people to these functions for future planning.

Chad relied on his training, experience, and knowledge of how to apply these in an unprecedented event. While knowledge and experience with implementing the ICS and conducting risk management can be applied in any incident, he maintained vigilance, not merely relying on certifications or qualifications. Chad asserted that "Every piece of training across multiple disciplines was extremely helpful to have during this crisis. The first responder community needs to have a broader understanding and perspective on the multitude of disciplines, not just their specific discipline to better integrate in the event of a disaster of this magnitude to make for a more efficient and effective response." Multi-discipline courses and exercises enable the integration of these disciplines to illuminate the macro perspective. This is necessary experience for large-scale, multi-agency, multidiscipline response operations. COVID-19 was not Chad's field of expertise, but he was able to relate it to other incidents such as Hazmat for frame of reference.

As a state resource, Chad was able to leverage his relationships across the state and a broader perspective to assist the county. For example, he facilitated meetings and cooperative agreements with adjoining counties to share information and redistribute resources such as testing kits and PPE. Through the County Health Department, contracts were established for testing with ambulance providers for public safety responders. As capacity allowed, this was expanded to the public. Having an established network of relationships before the incident was an essential element of success.

There were two leadership skills that Chad identified as essential to successfully navigating this crisis: clarity on leader's intent and transparency with the team and the public.

Clarity on leader's intent enabled his team to define the task, purpose and end state which created buy-in. This also laid the foundation for the long-term strategy as the end was not clearly in sight. Relying on checklists and regular reevaluation, ensured a consistent approach, reduced stress, and projected confidence among the team.

Transparency with his team required the humility to admit when he did not know something as an IC. This helped establish trust. Being transparent enabled him to maintain calm in the middle of a stressful situation. Keeping the public informed was essential and necessary. This ensured accountability and provided an opportunity to show that decisions were made based on science and the data.

Aspiring leaders should ensure that they have foundational knowledge in their core discipline but also focus on topics that are not mainstream such as public/private integration and international public leadership. Leaders need exposure to other topics: morale, leader's intent, application of their knowledge, establishing SMART objectives, team buy-in, and communicating the "why" behind the incident and response operations. While common terminology is an essential element of the National Incident Management System (NIMS), responders need to use layman's terms. Some responders have a tendency to get too technical when serving in a multidiscipline environment; they get drawn into the exactness of the ICS doctrine, sometimes preventing the ability of the incident to be efficient. Although technically precise, colleagues may get lost in the complexity and details which results in ineffective communication. Clear, plain language is extremely beneficial in rapidly changing environments.

To set the conditions for success during a crisis, Chad recommends communicating the leader's intent and then refining the mission through the development of objectives, strategy, and implementing tactics that look to favorably change the outcome – this must be defined. Prior to the incident, response organizations must prepare and equip to establish their response capabilities. Leaders must also rely on and apply valid data and science to support decision-making to the greatest extent practical. Finally, leaders, especially in a multi-jurisdictional operation, must establish trust with their local leadership counterparts; they understand the local and operational aspects of the incident. This entails asking questions and building understanding; ego among leaders in the ICS is a point of failure.

10.7 Major General (Ret) Dana Pittard: Leading the Campaign against ISIS in Iraq

This section is based on an interview that I conducted with General Pittard on April 12, 2021.

Major General (retired) Dana J.H. Pittard served for 34 years in the United States Army. While serving as the Deputy Commanding General for Operations for the Army Component in Central

Command, he was working in Jordan and Kuwait as the Islamic State in Iraq and the Levant (ISIS, Daesh) was taking over large portions of Syria and Iraq. In January 2014, ISIS had taken control of Fallujah, Iraq. By June 9, 2014, ISIS had taken control of the city of Mosul. As they proceeded down the Tigris River valley, they defeated five Iraqi divisions and four police divisions. These events triggered alarms as they seemed unstoppable. While visiting troops in Egypt, General Pittard was called back to Camp Arifjan, Kuwait, and charged with the task of establishing a command-and-control headquarters in Iraq, assisting the US embassy in case of a potential evacuation, and advising the Iraqi leadership.

Within 72 hours, General Pittard arrived in Iraq on Jun 23, two weeks after ISIS had taken over Mosul. With strict political limitations on operations and the assassination of Ambassador J. Christopher Stevens in Benghazi still fresh in everyone's memory, General Pittard assumed command as the Joint Forces Land Component Commander–Iraq and successfully led 300 troops and a coalition of forces over the next five months through a pivotal crisis as ISIS advanced toward Baghdad.

During his first 48 hours in command, he assessed the situation by synthesizing his personal experience in Iraq, his prior training, his knowledge of military history, and input from the Iraqi leadership. Based on a high degree of situational awareness, he made a somewhat controversial decision believing that Baghdad could hold off ISIS and defend itself. The Iraqi President then made a statement to his country to defend Baghdad. General Pittard also recommended striking ISIS and developed a campaign plan during this assessment period while the Administration figured out their plan for responding to ISIS. General Pittard worked closely with colleagues to influence decisions with his insight and input.

Having been to every province in Iraq, General Pittard was very familiar with the people and factions. He had established trust with Iraqi leaders during a prior deployment as part of the US–Iraqi transition team. This enabled him to ask good questions about the local situation: he knew what to ask and he had built a context for the situation prior to the campaign. He utilized various systems to gather intelligence (e.g., technology, Human intelligence (HUMINT) from the Iraqi army and tribal relationships) but needed to compile and filter this information to determine credibility and reliability. For example, the US Defense Attaché claimed that ISIS was unstoppable. But General Pittard knew that this was not the case because at some point, the enemy would slow their operations tempo, and he could disrupt their decision cycle and ops tempo. This would eventually enable the coalition to control the momentum tactically and operationally. Nonetheless, leveraging the whole of government (e.g., Treasury, NSA, FBI, etc.) became very important to enhancing situational awareness.

> Operations tempo is the speed and intensity that operations are conducted to include shift periods, shift changes, breaks in battle, resupply. This is relative to the speed and intensity of unfolding events in the operational environment. Maintaining an operations tempo ensures proactive management of the crisis and prevents leaders from being forced to react to the environment. Controlling the operations tempo is an advantage during a military operation.

Another key decision entailed maximizing intelligence, surveillance, and reconnaissance (ISR) resources to monitor ISIS, learning their tactical movements, and capitalizing on superior intelligence. ISR resources were not easily obtained, so General Pittard leveraged his relationships across organizations and the coalition to ensure that his requests were prioritized and acquired the analysts needed to process the intelligence.

While he did not always have complete information, he did not shy away from making decisions. He used the best information available to keep moving forward. Many senior leaders were uncomfortable with this as a risk averse culture dominated the Administration and some senior military leaders. General Pittard's philosophy was to keep moving forward by making decisions that aligned with the objectives and made "wrong decisions" right. He said, "Leaders need to be OK with making mistakes but they need to learn from them. Leaders cannot be afraid to make decisions to achieve the objectives." This would lead to gaining and maintaining momentum on the battlefield. Leaders need to be allowed to make mistakes during training and exercises, otherwise they will never want to take risks. The leader's supervisor needs to underwrite those mistakes to allow growth. General Pittard believes that it is in his DNA to take calculated risks to achieve the desired effect.

ISIS was very disciplined compared with other militias. Their mode of operation entailed taking over people and territories, then taxing the people to fund their expansion and military operations. Then ISIS made a strategic blunder; in August 2014, they attacked the Kurds in Irbil and started killing hostages. The Kurds requested assistance. General Pittard worked with senior military and Administration leaders to obtain permission to strike ISIS in order to defend the Kurds. While fighters stopped slightly outside of the approved targeting perimeter, General Pittard was faced with another critical decision: delay striking and miss the opportunity to catch the enemy by surprise or order the air strike. He knew that ISIS was training 400 people per month and that in order to overcome this, the coalition would need to kill more than 400 people per month. On August 8, 2014, he ordered air strikes on massive formations of ISIS fighters marching toward Irbil.

Political restrictions and micromanagement of air strikes from the Administration did not hamper General Pittard and his approach to navigating this crisis with agility and precision. While he was required to obtain permission to strike limited, specific targets rather than anywhere in Iraq, he would submit targeting slides containing specific, strategic locations throughout Iraq. Since the slides were devoid of an Iraq-wide orientation, he was able to comply with the restrictions while having the ability to strike where needed throughout Iraq.

Another major leadership challenge was the need to build a coalition and trust among opposing factions within Iraq in order to fight a common enemy. General Pittard's familiarity and experience with these groups enabled him to work with them and bridge the trust gaps that existed in order to build a unified coalition against ISIS. Iraqis and Kurds opposed each other. Iraqis were split among Sunni and Shi'a Muslims with Shi'a groups backed and connected with Iran. Within each sect, there were multiple militias with their own caveats to participation. The Western coalition also had their constraints and individual agendas. The degraded Iraqi military had lost the will to fight; however, the Iraqi Counter Terrorism Service (CTS) had maintained their training and will to fight. So, he started with the CTS and the Kurds. In the wake of ISIS seizing control of the Mosul dam, General Pittard formed a coalition between the CTS and Kurdish Peshmerga by negotiating with and influencing both sides to work together. He leveraged opportunities to support humanitarian missions to build the coalition of the unwilling. He facilitated finding common ground against a common enemy by keeping the various factions focused on ISIS. This led to joint operations with small wins such as taking back the dam which grew to bigger wins on the battlefield and helped to build confidence and trust. Ultimately, this momentum led to a pivotal moment in the fight against ISIS where the coalition was able to choose the time and location of attacks. By October 2014, there were 15 battle fronts, the operational tempo was increasing, and the campaign blunted ISIS's ability to conduct major operations.

While General Pittard had been battle tested through multiple combat tours, this campaign was different. It was ambiguous and he was not in control of all resources. This was a source of stress, but he was able to overcome the desire for control by focusing on achieving effect, not control. He leveraged his knowledge of history, experience, and training to build the coalition and focused them on defeating the common enemy.

After Iraqi victories, the Kurds would kill hostages. This weighed heavy on his mind, but he knew that defeating ISIS would eventually lead to cessation of this practice.

Another stressor was the lack of sleep. This was caused by two factors: (1) the Iraqis tend to work later into the night and start their day later in the morning, while the Western forces would start their day earlier and end their day sooner; (2) the restrictions on air strikes required approval from two General officers, so to ensure that opportunities were not missed, he would spend many nights at the Baghdad airport so that he could awake and review the strike packages in a timely manner.

To help manage these stressors, General Pittard would work out a lot. In fact, his best ideas came while swimming laps. He was able to think clearly and identify patterns such as how ISIS leaders would visit their troops in the field and then the coalition was able to exploit these patterns. Additionally, he managed to maintain at least six hours of sleep per night. This balance of work, exercise, and sleep helped General Pittard to remain mentally fit in a constantly stressful and demanding situation.

General Pittard identified the following essential leadership skills that enable a leader to successfully navigate a crisis. First, know yourself, specifically your strengths and weaknesses. Personality profiles such as Myers-Briggs may be helpful in knowing yourself. Second, know others around you. This is essential in building trust and relationships. This also leads to being able to surround yourself with people that are different from yourself and recognizing the talent around you. As you get to know others, you will also care for them. Third, build personal courage to do the right thing and take risks. Be willing to stand up to other leaders when necessary and communicate the situation, especially when facing adversity or sharing unpopular news; tell it like it is. Finally, stay fit. This is the key to resilient leadership. By taking care of yourself, eating healthy, getting enough sleep, and working out, you will be able to stay focused on the objective during a crisis.

Developing these skills requires building experience as a leader and becoming a lifelong learner. Leaders must commit to personal study, training, and leverage their intuition. Do not discount intuition.

Setting conditions for success during a crisis requires understanding the parameters and making decisions that result in the desired effect. In many cases, doing something is better than nothing. Crises will often create ambiguity. A crisis leader needs to be able to move forward through decision-making and find innovative ways to accomplish objectives with limited resources and political constraints. Surround yourself with the talent and people. "When you have the ball, you run with it. The greater your authority, the more responsibility you have to serve. So, exercise humble leadership."

Application

Answer the following questions for each profile:

1) What attributes helped the leader achieve successful outcomes?
2) Describe the stressors that they faced and how they managed stress?
3) What were the key decision points in the case study?
4) What resources did the leader use to increase their situational awareness? Which resources were most or least useful?
5) Identify the external stakeholders and how the leader influenced these stakeholders.
6) How did these external parties attempt to influence decision-making and assess how the leader handled these influences?

Identify a circumstance where external parties were attempting to influence a decision you were making. How did you handle the situation? What techniques can you learn from the profiled leader that could have improved your handling of the external influencer?

Political influence on an incident is common and crisis leaders need to manage the incident and address the political influences before the political event becomes its own crisis. What can we learn from our case studies to improve how we address political influence during a crisis?

References

1 National Oceanic and Atmospheric Administration (2019). 2018's billion dollar disasters in context. https://www.climate.gov/news-features/blogs/beyond-data/2018s-billion-dollar-disasters-context#:~:text=The%20combined%20costs%20of%20the,adjusted%20to%20January%202019%20 dollars.&text=The%20annual%20cost%20average%20for,record%20(1980%E2%80%932018) (accessed January 8, 2022).

2 Insurance.com (2018). Study shows half of homeowners don't understand liability insurance. https://www.insurance.com/press-room/study-homeowners-liability-insurance-2018 (accessed May 24, 2021).

3 Prosperity Now (2016). Financial assets and income: asset poverty rates. https://scorecard. prosperitynow.org/data-by-https://scorecard.prosperitynow.org/data-by-issue#finance/outcome/ asset-poverty-rateissue#finance/outcome/asset-poverty-rate (accessed May 24, 2021).

4 Haveman, R. and Wolff, E. (2004). The concept and measurement of asset poverty: levels, trends and composition for the U.S., 1983–2001. *Journal of Economic Inequality* 2: 145–169. doi: 10.1007/ s10888-004-4387-3.

5 The Last House Standing (2019). Documentary film. https://www.thelasthousestanding.org (accessed May 24, 2021).

6 McMaster, H.R. (2021). *Battlegrounds: The Fight to Defend the Free World*. New York: HarperCollins.

7 McMaster, H.R. (2017). *Dereliction of Duty: Lyndon Johnson, Robert McNamara, the Joint Chiefs of Staff, and the Lies that Led to Vietnam*. New York: HarperPerennial.

8 Clausewitz, C. (1960). *Principles of War* (trans. H. Gatzke). Harrisburg, PA: Stackpole Books.

9 Complexity Theory – Key Concepts (2019). https://www.youtube.com/watch?v=hLXIJF5ytpM (accessed May 24, 2021).

10 Interaction Design Foundation (2021). What is design thinking and why is it so popular? https:// www.interaction-design.org/literature/article/what-is-design-thinking-and-why-is-it-so-popular (accessed May 24, 2021).

Further Reading

1 Buffaloculture.com

2 www.c12group.com

3 Pittard, D. and Bryant, W. (2019). *Hunting the Caliphate: America's War on ISIS and the Dawn of the Strike Cell*. New York: Post Hill Press.

11

Attributes of a Crisis Leader

After reading the profiles of our crisis leaders in Chapter 10, you may have noticed some commonalities that led to their readiness to succeed. The goal of this chapter is to bring those attributes into focus in order to help us recognize a crisis-ready leader, assess whether a person has what it takes to become a crisis-ready leader, and to formulate the next steps in becoming one.

Learning Objectives

After reading this chapter, you will be able to:

- Identify the character traits, knowledge, skills, and abilities of a crisis-ready leader.
- Assess whether a person has these attributes.
- Formulate the next steps in developing these attributes.

The scope of this chapter is to focus on attributes relevant to crisis leadership. There will undoubtedly be some overlap with general leadership attributes, but since there are many books written on leadership, we are going to confine the scope of this chapter specifically to those attributes necessary for a crisis leader. Think of this list as a subset of leadership attributes with heavier emphasis on those needed to successfully navigate a crisis. A good crisis leader is likely a good leader, but not every good leader is ready to lead through a crisis. Let us dive into the attributes needed for crisis leadership.

11.1 Essential Character Traits

As I interviewed each of the crisis leaders for our profiles, I was reminded by several of them that it takes more than just knowledge, skills, and experience to lead through a crisis successfully. Character matters. As we explore each of these character traits, think of them on a scale or spectrum rather than thinking that a person either has it or does not. No one is perfect. You are reading this book to grow and develop so that you are prepared to lead in time of crisis. Being self-aware of where you fall on the spectrum of each character trait creates the opportunity to work on those areas needing improvement. Start by asking others – a supervisor, spouse, coworker, accountability partner, pastor, or friend – to help you develop in these areas.

Crisis-ready Leadership: Building Resilient Organizations and Communities, First Edition. Bob Campbell, PE.
© 2023 John Wiley & Sons, Inc. Published 2023 by John Wiley & Sons, Inc.

11.1.1 Humility

> "Humility is not thinking less of yourself but thinking of yourself less." C.S. Lewis

All of our crisis leaders referenced some aspect of humility and several of them directly stated that humility was a necessary trait. Chad Hawkins identified the need to admit when he was wrong and share information with others on his team in a transparent manner. This built trust and quickly strengthened the bond among team members as he assisted a locality as the incident commander during the early stages of COVID-19. While he was in charge according to the organizational chart, he treated others with respect, listened, learned, and led. H.R. McMaster discussed the importance of strategic empathy when working alongside Iraqis and Turkomen in Tal Afar.

> Strategic Empathy is a deep examination of the emotional, cultural, and political drivers which help us understand another's drivers and constraints. When a person is under stress or facing a crisis, their behavior reveals what matters most to them. This is not only important to understand for military leaders who are battling a dynamic enemy, but also for civilian leaders responding to and recovering from a natural disaster that upends a community, or civil unrest where protesters become violent.

Humility moves the focus of our attention from ourselves to others. Are we more inclined to be interested or interesting? A humble leader focuses on others, rather than notoriety or gain for the self. The people that you lead can detect this based on what you say and what you do. Do you speak more than you listen? Do you exercise empathy with others and truly seek to understand different perspectives, or do you know all the answers and therefore you do not need to listen?

Humility attracts others to a leader because they feel heard, cared for, and empowered to contribute to the team. A good leader can recognize the contributions of others on the team and leverage their contributions for the good of the team. Putting others first and recognizing their success encourages team members to perform well. During a crisis, there is not a lot of time to get to know each other, build trust and relationship, or figure things out. A crisis leader needs to bring together a multitude of team members, even some that they may never have worked with, and influence everyone to work for the same mission and objectives. This can be challenging when some team members will bring their own agenda to the incident. A humble leader listens, learns, and leads with a sincere interest in the team members. As Chad Hawkins stated, "[humility] leads others to calm" during a chaotic situation.

11.1.2 Integrity

Integrity is a key component in building trust. Patrick Lencioni, author of *The Advantage* [1], conceived how trust enables teams to engage in healthy conflict, brainstorming, and evaluation of alternatives. These contributions from experts engaged in debating the merits of alternatives which lead to buy-in among team members. Once the team has bought in to the decision, they become accountable. When they are accountable, the team achieves results. The foundation is trust.

Trust is like a bank account. Team members make deposits by demonstrating competence, respecting others, and doing what they say they are going to do. When these events occur, trust is built. But when team members fail to deliver, trust diminishes. Integrity is derived from the Latin word *Integer* which means whole or completed. Integrity has several meanings to include – adherence to moral values, soundness, and completeness. Some would describe it as doing the right thing even when no one is looking.

Each of these aspects of integrity are necessary to lead people through a crisis when split-second decisions need to be made and the public and your team members are trusting that you will make the right decision. If a leader has a track record of making bad decisions, this should be a red flag indicating that this person might not be fit for leadership position during a crisis. On the flip side, if the leader has exercised sound judgment in decision-making which has led to positive outcomes, then this should be a green flag indicating crisis-leader potential. Take some time to evaluate quality of previous decisions and the associated outcomes. Are you living a life aligned with a good moral foundation? Can you define that good moral foundation? Many organizations adopt core values. These are the values that are timeless and unwavering despite the circumstances. This may be a good place to start in defining those uncompromisable values that you adhere to regardless of the situation.

In addition to the moral aspect of integrity, completeness and wholeness are necessary for one to rely on someone else. At the onset of this section, I highlighted how integrity, doing what you say you are going to do, establishes trust. When our words, beliefs, and actions align, we demonstrate integrity. However, when we say one thing and do another, our integrity takes a hit. It can be hard to recover our integrity without the humility to admit the wrong. People do not want to follow leaders that they cannot trust. Unfortunately, in our fast-paced world, it is easy to forget to fulfill promises made. While being late for a meeting, forgetting to return an email, or completing a task assigned at the last meeting may seem like minor oversights, these omissions can damage a leader's integrity and result in a depleted trust account.

To keep your integrity in check, it is important for leaders to create reliable systems to keep track of commitments made, follow up appropriately to verify completion, and continuously improve. Frank Patterson stated, "The keys to setting conditions for success start before a crisis with establishing integrity."

11.1.3 People Smarts

In his book, *The Ideal Team Player* [2], Patrick Lencioni identified the three attributes of the ideal team player as humble, hungry, and smart. He underscored that smart did not merely imply knowledge, but moreover an understanding of self and other people. Good leaders possess a high degree of self-awareness and situational awareness about the people they work with. They know who they can trust, the strengths and weaknesses of those on their team, and their personality traits. This begins with caring enough about others to get to know them well. Some organizations utilize personality assessments such as Myers-Briggs, DISC, etc. to help team members get to know themselves and others on their team. Dana Pittard was quick to share his Myers-Briggs profile with me and point out the profiles of those he needed on this senior leadership team to complement his personality, strengths, and weaknesses. He also explained that there are some profiles that will clash and how this could be problematic during a crisis. H.R. McMaster underscored the importance of listening and exercising strategic empathy; this enables a leader to understand the various stakeholder perspectives, motivations, and concerns. The key is understanding yourself and how you work well (or do not work well) with others.

In my company, I utilize The Hiring Suite personality profile. This eight-point assessment helps identify the personality of candidates and then produces a Team Master report that compares each personality trait between a supervisor and their direct reports. This report highlights areas of synergy as well as potential areas of conflict or misalignment. It also provides advice on how to work with those personality traits that might clash so that coworkers better understand themselves and others. We find that this is a helpful tool in ensuring diversity among our workforce while preempting conflicts rooted in different personalities.

Regardless of how you increase your people smarts, recognizing the need to possess people smarts before a crisis can help leaders prepare. Because a crisis can be fast-paced, there may not be time to figure this out in the moment. Leaders who lack in this area should uncover it before the crisis, work to improve, but surround themselves with someone on their leadership team that excels in this area and can help read the people situation.

Additionally, our culture values logic and reasoning over intuition. While we need to keep our biases in check for many reasons, this attempt may also squash intuition. While intuition lacks the objectivity of logic and reason, some team members possess genuine intuition, meaning they have a good sense or instinctive knowledge about something. As leaders strive for decisions based on science and rational thinking, there is some value in tempering the obsession for data with professional judgment and intuition. Dana Pittard advised crisis leaders, "leverage intuition ... do not discount it." Perhaps his intuition played a role in identifying patterns among ISIS leaders that he later exploited.

11.1.4 Moral Courage Builds Confidence

I asked Dana Pittard about the risks he assumed as he made life-and-death decisions pursuing ISIS in Iraq while managing political, media, and international adversity. He stated that leaders need to "build personal courage to do the right thing and take risks. Be willing to stand up to other leaders when necessary and communicate the situation, especially when facing adversity or sharing unpopular news; tell it like it is." Too many leaders have taken the risk-averse approach to decision-making based on the potential personal consequences of making a mistake, perhaps a career-ending mistake for senior military and civilian leaders in government. These factors have the potential to influence a leader's decision-making process.

> **Moral Courage** is the willingness to act rightly in the face of popular opposition, shame, discouragement, or personal loss.

Having the personal and moral courage to evaluate a situation and make the best decision without bias from the opposition and concern over personal loss is essential for leaders who are navigating a crisis. There will likely be contingents that oppose your decisions. If it is the right decision, you will gain and build confidence from those you lead. If it is the wrong decision, Frank Patterson advises, "People are looking to leaders to make decisions; this is the primary expectation. Do not get paralyzed by analyzing all of the consequences or attempting to make the perfect decision. The wrong decision is 'not making a decision.' Do something." An advancement toward accomplishing objectives is a move in the right direction. Leaders must have the courage to move forward.

Like trust, confidence is gained based on achieving effective actions. But prior to this stage, a leader needs to project confidence. An indecisive leader will draw concern and reluctance from their team members. Projecting confidence is not arrogance and self-righteousness, but rather is based on one's readiness to lead, clarity in mission, and clear plan. Utilizing checklists and procedures can demonstrate confidence in the process and advance the response process. In chaotic, uncertain circumstances, relying on established plans and procedures can help create momentum and confidence. A leader's moral courage is needed to make the right decision in the face of opposition. Derrick Vick faced an impossible situation as cash flow diminished, the bank called Freedom's loans, and vendors demanded payments. It would have been easy to give up and file for bankruptcy. Despite the opposition and personal risk, Derrick marched onward exercising moral courage with a systematic and intentional discipline of focusing his efforts on increasing cash flow with new projects that were profitable, metering payables to satisfy vendors, and retaining employees needed to perform the work. He was able to maintain confidence and joy in a time of crisis.

11.2 Knowledge

Knowledge is gained through study and experience. All of our profiled leaders emphasized the importance of lifelong learning from reading, studying history, learning lessons from others, and gaining experience in their field.

11.2.1 Self-awareness

To be a good leader, one needs a high degree of self-awareness of whether they have the traits required, skill set, and ability to lead in a crisis. A good leader understands their strengths and weaknesses and inspires others to follow them. As discussed above in People Smarts, self-awareness is essential knowledge for any leader to uncover their strengths, weaknesses, and how they engage with others in an organization. In a time and resource-constrained environment, it is important to know what you know and do not know. Using Rumsfeld's Known Knowns model illustrated in Figure 11.1, leaders need to increase their awareness of what they know by acting to fill the knowledge gap with information needed to support decision-making. This applies to both self and the situation. Our goal is to minimize blind spots about ourselves by becoming more self-aware in the same way that we

Figure 11.1 Rumsfeld's "Known Knowns".

need to minimize blind spots about the situation so that we are more informed and can make better decisions. Leaders need to also leverage their intuition, or "unknown knowns," as a resource. As we become more aware, about ourselves, our team, and our situation, we minimize the blind spots that prevent us from fully knowing ourselves so that we can fully leverage the strengths of others to compensate for our weaknesses. A leader needs to uncover this information before the crisis.

Leaders should leverage their strengths during a crisis and leverage the strengths of others to mitigate their weaknesses. If a leader does not recognize their weaknesses, they will not do a good job of integrating others into these roles or deferring to them when needed. A good leader does not have to be good at everything, but a good leader needs to be able to build a strong team and facilitate their progress.

During the process of self-assessment, a leader may uncover that they will not make a good crisis leader. Not everyone can handle the stress or possesses the character, knowledge, skills, or ability to succeed as a crisis leader. Being honest with your leadership regarding your abilities and limitations could prevent a crisis from becoming a disaster.

11.2.2 The Plan

Organizations have plans for strategy, disasters, combat, business continuity, operations, mitigation, response, and recovery – and the list goes on. Knowledge and familiarity with the plan can mean the difference between a short and long recovery period. Some have posited that "plans are worthless, but planning is everything." I recall the Joplin, MO, tornado that ripped apart a hospital. In the aftermath of this incident, the medical responders reported that the plan blew away, but the planning process was valuable in informing them on what to do. Participating in the planning process certainly informs those involved, but not everyone is involved in the planning process. This did not stop Brock Long from reaching out and inviting external stakeholders to provide input into FEMA's strategic plan. Thousands of comments poured into FEMA and shaped the three strategic goals for the organization and the US.

Plans are essential tools for communicating key information on how an organization is going to accomplish some mission or objective, sometimes during a contingency. Nonetheless, plans are not immutable. Good plans include information on how it will be reviewed, maintained, updated, and revised, and when these reviews will occur. Each of the leaders interviewed cited an existing plan or the process of creating a plan to address the crisis. There is a place for both deliberate, pre-incident planning and crisis-action planning. Leaders should be familiar with the deliberate plans either through review, training, or participation in an exercise, but leaders should also be familiar with the crisis action planning process as they may need to update or create a plan during a crisis.

Having a common plan in place before the crisis is a good starting point when later adapting plans to unique circumstances. COVID-19 is a good example. Most communities, health departments, and states had a pandemic response plan on the shelf for the last decade. However, the uncertain nature of the virus (i.e., its transmissibility, virulence, method of transmission, and incubation period) disrupted many plans and required additional actions. As the incident commander for his incident management team, Chad Hawkins identified that the county health department had a plan in place but needed an incident action plan. So, he utilized the planning section of the ICS to establish commander's intent, incident objectives, and strategies to create steady-state operations at the local level. This planning process led to the early development of contact tracing, mass vaccination, and mobile testing plans.

In the wake of the West Fertilizer Company explosion, Frank Patterson was able to rely on local plans to keep focused and manage the effects of stress during the response and recovery. However, one of his key decisions was to deviate from a mass casualty plan to transport casualties to certain medical centers. He made this decision because of adverse weather conditions which would eliminate air transportation, so he maximized his air transportation capacity to the farthest hospital considering each hospital's treatment capacity while he had it available, and then utilized the nearest hospital when only ground transportation was available as the frontal system passed.

As Dana Pittard arrived in Iraq, one-third of the US embassy had been evacuated as ISIS moved swiftly down the Tigris Valley. During his assessment of the situation, he directed strikes against ISIS and the development of a campaign plan while the Administration evaluated the situation. While this was a unique situation, a plan was necessary to coordinate the whole of government, Iraqi, and coalition forces.

Plans provide an organization with a roadmap for handling some crisis before, during, and/or after the event. The planning process typically begins with a risk assessment to inform the plan's assumptions and priorities. It spans the duration of the incident and may establish decision criteria and decision points. Leaders should become familiar with their plans and the organizations capabilities through exercises that verify the plan.

11.2.3   Situational Awareness with Leading Indicators

Situational Awareness informs decision-making. Decision quality can be improved with good situational awareness. As discussed in Chapter 5, situational awareness occurs when raw data inputs are compiled to make sense of a situation (i.e., comprehension), and a projection concerning the future outcome can be made. This may occur mentally or with the aid of models or decision-support tools. Understanding whether inputs are lagging or leading indicators can improve decision-making during a crisis.

Derrick Vick of Freedom Industries quickly determined that the best leading indicator of business survival was projected cash flow. Using routine expense data and future project cost and income projections, in conjunction with a safety factor, he was able to predict cash flow and make decisions and changes to positively impact and ultimately reverse the negative outcome. This objective approach illustrates how leading indicators enhance situational awareness and can lead to improved outcomes. Had Derrick made decisions based on a lagging indicator such as his profit and loss statement or the balance sheet, Freedom would have gone bankrupt.

Brock Long, former FEMA administrator, identified that asset poverty is a major contributing factor to resiliency. He suggested that communities should use the average credit rating as a leading indicator for civil unrest, divorce, disasters, food deserts, and underinsurance.

General Pittard recognized ISIS leadership patterns and was able to capitalize on these leading indicators in order to target ISIS leadership.

Frank Patterson incorporated the weather forecast to gain greater situational awareness regarding patient transport options to local hospitals and the viability of ground and air transport with impending weather.

Col McMaster utilized HUMINT conducted in raids leading up to the major operation to gain insight into AQI's plans and organizational structure.

It is incumbent on leaders to know the difference between lagging and leading indicators so that their situational awareness is informed and shaped in a way that enables projecting into the future. This becomes more difficult in a complex adaptive system that evolves based on human inputs. Whatever your context and crisis, seek out the leading indicators that you need to rely for shaping the outcomes.

11.2.4 Multidisciplinary Knowledge

Chad Hawkins and Frank Patterson, both career firefighters, acknowledged the importance of understanding the capabilities and resources from other disciplines and the need to learn from others. While leaders should ensure that they have foundational knowledge in their core discipline, they should also focus on learning topics that are not mainstream or within their discipline, such as: public–private integration and international public leadership. Learning alongside leaders from other disciplines will create a context for broader perspectives and discussions. While each discipline has a specific body of knowledge and ways in which they approach problems, this diversity of thought facilitates a collaborative, multidisciplinary environment.

Chad Hawkins' experience during the COVID-19 pandemic illustrated the importance of learning from other disciplines. In his position at the state fire marshal's office, he deployed for over a month to assist a county health department in the early stages of the pandemic. He was not a public health expert, but he understood incident command, unified command, the importance of private–public collaboration, and how to leverage his network in other counties and state agencies. Having worked and led in multidiscipline environments, Chad was familiar with the health department's mission and role in responding to epidemics, but this was an atypical response. As he continued to lead the work on mass testing and vaccination plans, he applied his experience in working outside of his discipline as a fire fighter.

Generals McMaster and Pittard also highlighted the importance of learning history, learning from the experience of others through lessons learned, gaining interdisciplinary perspectives, and working beyond their discipline. Broader knowledge of the peoples and cultures in Iraq enabled them to connect and build trust in a short period of time. Understanding the capabilities of their partners from other military services, coalition forces, and other government agencies enabled them to marshal disparate groups and organizations into a unified operation.

As we explore the career pathway of a crisis-ready leader, perhaps this framework will help:

- Early career: Discipline-specific knowledge and experience
- Mid-career: Leadership foundation and multi-disciplinary learning
- Late-career: Leading multi-disciplinary groups within an organization
- Executive-level leader: Leading coalition of organizations

Most of a leader's later career will be engaging with other disciplines and organizations. To prepare for this pathway, leaders should be engaged in learning the basics of other disciplines and collaborating in a multidisciplinary environment. As we have explored throughout the book, crisis leaders will inevitably depend on external organizations to provide resources in the response and recovery phases of a crisis.

11.3 Skills

Skill is the demonstrated ability to perform a task by applying one's knowledge, experience, training, and aptitude and can be measured in how well the task is performed.

11.3.1 Communication

Communication is frequently cited as a major opportunity to improve (i.e., a problem) in after-action reports. Why is communication such an issue? Communication is not unilateral or theory, but rather it is multilateral and practical. It requires a sender, message, media, and receiver. Regardless of whether you are the sender or the receiver, you are engaging in communication.

To be effective in communication, one must be proficient at receiving the message (i.e., listening, reading, knowing which media to utilize in receiving the message, and comprehending the message). Transferring this onus to the sender puts the receiver at a disadvantage. Effective communicators understand how to be a good listener or receiver, and how to enhance clarity by asking the right questions. Being an effective communicator starts with being an effective receiver.

- Know what sources of information are reliable and consistent and how to gain access.
- Understand the sender, their context, and any barriers to communication such as jargon.
- Bring clarity by summarizing the message and asking good clarifying questions.
- Be attentive, empathetic, and interested.

The media itself is a component in communication and can affect the message and the receiver. There are multiple media that can be used for disseminating a message. These include: local broadcast television, cable television, radio, emergency alert systems (EAS), press conferences, written press releases, various social media platforms, email, text messages, phone calls, ICS forms, written notices, etc. Accessibility, timeliness, length of message, and importance affect the communication. For example, when developing a communications strategy with vulnerable populations, the sender should consider their audience and accessibility. During the COVID-19 pandemic many public and private organizations shifted from in-person to online services. However, for many elderly constituents, utilizing smartphone apps through the internet was a challenge due to lack of equipment, internet savvy, and accessibility. Impoverished populations may lack a television or smartphone with EAS notifications. With instant access to information, many people are able to access information about a crisis prior to official communications due to the time it takes to coordinate and disseminate information. Response organizations may rely on shift change briefings, leader's intent meetings, ICS forms posted in a common operating picture, or direct communications. Frank Patterson shared how he would hold update briefings for responders on the back of a trailer just to keep them informed with consistent and current information. Selecting the right media and adapting the message according to the media being utilized are important to the communications process.

The message is the third component in communication. The message may be written, graphical, or verbal. As diversity grows with the diaspora of people throughout the world, language barriers are increasing. Ensuring that emergency messages are available in the language of constituents ensures that the message is accessible. Recognizing the literacy level of the population may also lead to messaging graphically in addition to written or verbal form. When communicating in person or through video, recognize that only one-third of our communication is verbal while the other two-thirds is comprised of tone and non-verbal communication. In a technology-rich environment, it can be easy to misunderstand a verbal message devoid of these other forms of communication. Leaders need to bridge these gaps. Effective messages answer basic questions such as:

- Who is the message from and who is the message intended for?
- What is the situation?
- When did the situation occur and when is the message disseminated? Is it still relevant?
- Where did the situation occur? Does it apply to me?
- Why is this message being sent? Is it informational or actionable? What action needs to be taken?

The sender is the final component in communication. A leader needs to recognize that as the sender they have a responsibility to draft, review, or approve a consistent and clear message, disseminate it through the right media to ensure that the right recipients receive the message, and be responsive to the audience or receiver to ensure the message was understood. Utilizing some type of feedback loop helps to prevent miscommunication. Some best practices include:

- Utilize common and consistent terminology.
- Be concise and empathetic.
- Avoid jargon or risk comparisons.
- Be sincere, direct, and honest.
- Send timely, relevant, and actionable information.
- Be responsive.

During a crisis, there will likely be different audiences that require different messages. At its most basic level, communication will occur between direct reports and their supervisor in the ICS, the public safety team and the public, the public safety team and the media, the media and the public, and employer and employees. Communication will also likely occur in higher headquarters, at the state/federal agencies, and informally among responders.

Derrick Vick identified several constituents requiring frequent communication: his employees, his Board of Directors, and the bank. Communication was a key determinant in retaining and leading employees through a crisis.

Brock Long established multiple communication channels to solicit input for the update to FEMA's strategic plan. These included not only employees but state/local/territory/tribal, private sector, and public stakeholders. Additionally, he recommunicated the strategic plan with many audiences wherever he spoke, ensuring frequent and consistent communication.

Leaders need to assess the effectiveness of their communication and work to continuously improve upon their writing, speaking, and style to ensure maximum effectiveness.

11.3.2 Complex Problem-Solving and Design Thinking

Complex problem-solving begins with understanding the basics of complex theory and complex adaptive systems. The essence of this theory is that we live in a complex, interdependent, interactive, and evolving world that includes a great deal of uncertainty due to the various complex systems at work. A defensive situation where leaders are presiding over systems committed to working together to solve the common problem is less complex than an offensive situation where there is an active threat working against the response. A natural disaster may be an example of a defensive situation whereas a complex coordinated terrorist attack may be an example of an offensive situation. These examples illustrate that there is some spectrum of complexity. Nonetheless, leaders need to recognize that these complex systems are not linear or completely predictable. Cause and effect are non-linear. Understanding the systems at work, their inter-related nature, and their evolution can help leaders navigate through the uncertainty.

Design thinking is an iterative, non-linear process of empathizing with others, defining the problem, ideating on potential solutions, prototyping the solution, and then testing the solution. Each step in this process could lead to any other step in this process until the design yields a viable, effective solution. Trying to arrive at the perfect solution on the first attempt will only delay actionable response. As General Pittard noted, "doing something is better than nothing." Likewise, General McMaster cautioned against rushing to action before defining the problem. Acting to solve the wrong problem could waste limited, valuable resources, but taking too long to define a problem will result in problem evolution and solving the wrong problem. Defining the direction and momentum of the problem can help intercept the problem with the right solution at the right time. Crisis management is analogous to building the airplane while flying it.

11.3.3 Trust Building

Building trust requires the confluence of integrity, competence, and respect for others. While everyone has the ability to build trust, this only occurs through repeated actions. Trust is similar to an account where a leader makes deposits based on demonstrating integrity, competence, and respect for others. When a leader fails to demonstrate these, the account is depleted. A leader begins building trust long before the crisis. Integrity is demonstrated when leaders do what they say they are going to do, provide evidence that their beliefs are consistent with their actions, and show they are capable of making good, moral decisions. A leader demonstrates competence by doing their job correctly, applying their knowledge and experience to situations, and making sound decisions that achieve positive results. Finally, and most importantly, a leader must respect others. This begins when a leader establishes a relationship, listens, cares, and responds empathetically toward another person. When these are genuinely demonstrated, a leader builds trust. During a crisis, it is likely that a leader will have had the opportunity to build trust with those within their organization, but as more external parties become involved, the leader will need to quickly find ways to establish a relationship and build trust. Here are a few tips for that fast-paced crisis situation:

- Be personable, transparent, and vulnerable – take the initiative on these traits.
- You never have a chance to make another first impression, or so the saying goes. There are different approaches depending on the cultural context. In the West, slow down, look into their eyes, shake their hand, learn about them as a person, maintain interest, use their name while talking with them. Strive to identify a common bond.
- Listen first, then speak.
- Admit when you are wrong and apologize.

Building trust is essential for exercising influence. Without trust, people will not respect or heed a leader's input.

11.3.4 Decisiveness

Good leaders are decisive. Frank Patterson stated that people are looking to leaders to make decisions; this is the primary expectation. He cautioned against becoming paralyzed by analyzing all of the consequences or attempting to make the perfect decision. The wrong decision is "not making a decision. Do something." He went on to emphasize that leaders need to "be confident in decision-making." Decisions move the response forward by changing the situation; good decisions positively affect the situation. General McMaster commented that military commanders cannot be paralyzed by waiting for more information to increase certainty. The riskiest course of action is to do nothing as that allows the enemy to take the advantage. Leadership and decision-making are inextricably connected.

11.4 Additional Advice

Several of our profiled crisis leaders emphasized the importance of **surrounding yourself with competent advisors and leaders**. They encouraged leaders to search for people that are not only smarter than them but also have a compatible and complementary personality profile so that the leader can leverage their strengths to compensate for the leader's weaknesses. These concepts build well-functioning teams. As Chad Hawkins assisted the county health department in establishing the ICS and developing an **ops tempo**, he worked toward the ultimate goal of transferring responsibilities and roles back to the county. This required training and development of competencies among local staff filling roles in the ICS. Once this was achieved, his mission was complete. During crises, leaders may not get to choose their preferred A-Team for positions in the ICS. Varied levels of knowledge and experience will likely require on-the-job training. Brock Long was fortunate to choose several of his key team members at FEMA and he highlighted the important roles they played in supporting his leadership and then transitioning from his leadership after he departed. Derrick Vick shared how important it was for him to have multiple advisors from his Board and C12 Group to provide the accountability and objective feedback that he needed to succeed.

Maintain objectivity throughout the crisis. As Frank Patterson shared, "you did not cause the crisis, your role is to lead others through the crisis by achieving outcomes." He shared how at times he found it helpful to compartmentalize the situation and control the **ops tempo** by slowing down or pausing the incident. General Pittard shared how staying mentally and physically fit enabled him to clear his mind, and it was during these times that he was able to make strategic connections and generate the best ideas. Chad Hawkins emphasized the importance of leader's intent, establishing incident objectives, and monitoring performance toward these goals. Community Lifelines provide the structure within which a leader can objectively identify and prioritize actions through intermediate objectives and end-points. Throughout COVID-19, many leaders echoed the importance of facts and science while others allowed politics to drive decision-making. Establishing and communicating SMART goals (i.e., Specific Measurable, Achievable, Realistic, and Time-bound) either at the incident objective or intermediate objective level can ensure clarity of intent, alignment of resources, and a means for assessing progress and effectiveness.

Brock Long emphasized doing the right thing as "good leaders are rooted in love and service to other people." In order to know what the "right thing" is, leaders need moral clarity and objectivity. One good indicator of moral clarity is the degree to which a leader's actions serve others over self. In selecting a leader, one should have had an opportunity to observe a leader in action and to work together with them (or interview those that have worked with them). As a developing leader, it is important to check your motives and ensure adequate transparency and accountability.

Controlling the operations tempo (i.e., ops tempo) begins with establishing a battle rhythm. A battle rhythm is what the military refers to as the cycle of meetings, updates, and decision-making. A battle rhythm coincides with an operational period. As defined by the US Department of Defense, it is "A deliberate daily schedule of command, staff, and unit activities intended to maximize use of time and synchronize staff actions." (Joint Publication 1, DoD Dictionary of Military and Associated Terms, January 2021). In emergency management, under the ICS, the Planning-P and operational period establish an operations tempo. Businesses also utilize routine activities to update, inform, and make decisions. For example, in my business, we hold a weekly staff meeting among the executive team members to get in sync regarding operations, management, and business development activities for the week. We also utilize a monthly all-call to communicate with all our team members on various aspects of the business to include administrative information, policy reviews, operations updates, celebrate successes, and provide situational awareness on upcoming projects, priorities, and the financial condition of the company. During crisis periods, organizations may increase the frequency of some activities while prioritizing some over others.

Leaders control the ops tempo by first establishing an ops tempo. This may frequently align with the operational period, key events, or key decision-points or cycles. These cycles create alignment among the next tier of leaders by setting expectations for gathering and reporting key information, conducting key activities, and making decisions within certain timeframes. When the threat or hazard is unchanging (e.g., a bomb exploded and most of the response actions involve stabilizing and recovering from the incident), a leader has more control over the ops tempo. However, when the incident is evolving, such as changing weather patterns or an active threat, the crisis leader needs to anticipate and respond to new situations in order to control the ops tempo. Generals McMaster and Pittard demonstrated this by exploiting situational advantages on the battlefield which created opportunities for positive outcomes. While they certainly had a normal battle rhythm for conducting their operations, they exercised some flexibility in responding to new information and intelligence that led to a strategic advantage. They also sought to identify the enemy's ops tempo so that they could disrupt it. While not every leader faces an intelligent opponent, all leaders need to anticipate, recognize, identify, and act on advantageous situations in a timely manner. This might include leveraging weather opportunities as Frank Patterson did with transporting patients by air to the farthest hospital while weather permitted air transport, or leveraging strategic opportunities as Brock Long did with establishing the Community Lifelines to improve and streamline the ops tempo at FEMA.

Frank Patterson explained how he controlled the ops tempo by slowing the pace of the incident through compartmentalization (i.e., breaking down the incident into smaller, manageable issues). He cautioned against letting the magnitude of the incident overwhelm the leader. Chad Hawkins improved situational awareness during COVID-19 response through "operational pauses" so that the response team could reassess the leader's intent, and progress toward established strategic objectives and associated tactics. While the operational reality continues at one pace, the crisis leader needs to assess how and when to create space to break so that their team can rest, zoom out, reflect, process, and re-engage. During his battle with ISIS, General Pittard created this space for himself through exercise. Leaders need to create both an individual ops tempo as well as an organizational ops tempo.

11.5 Developing Crisis Leadership Skills

The application section of this chapter outlines several methods for assessing your readiness as a crisis leader and setting professional goals to further develop these skills. Those interviewed in the previous chapter highlighted how leaders can prepare for future crises. The following are a summary of these recommendations which also align with the attributes of a crisis leader:

- Recognize how your personality profile complements or contrasts with others that work with you.
- Embrace lifelong learning. Commit to personal study and training.
- Read history, lessons learned, after-action reports, and learn from others by asking the right questions – identify successes and failures in order to leverage these experiences and apply the lessons in the future. Gain strategic competencies and an understanding of contemporary challenges through the context of history.
- Establish foundational knowledge in your core discipline but also pursue learning from other disciplines, not just their knowledge base, but their culture and pattern of thinking.

Table 11.1 Attributes of a Crisis-ready Leader.

Character Traits	Knowledge	Skills and Abilities
● Humility ● Integrity ● People smarts ● Moral courage	● Self-awareness ● The plan ● Situational awareness with leading indicators ● Multi-disciplinary knowledge	● Communication ● Complex problem-solving and design thinking ● Trust building ● Decisiveness
Additional Advice		
● Surround yourself with competent leaders and advisors ● Control the ops tempo ● Maintain objectivity		

- Gain experience, book knowledge is insufficient. Find mentors who are willing to provide leadership opportunities where you can learn to navigate a crisis situation, but be willing to step aside if you become overwhelmed. Serve in positions where you can observe and learn from seasoned leaders. Likewise, invest in other leaders by creating opportunities for them to experience crisis leadership and decision-making.
- Surround yourself with other good leaders and smart people. Seek mentoring and advice.
- Trust but verify information. Ultimately, you are responsible for the decisions you make.
- Exercise a realistic span of control, i.e., no more than seven direct reports.
- Recognize that you do not have all the great ideas; create an environment where other experts can contribute to solving problems and creating solutions. Seek the best ideas and solutions.

Summary of Key Points

In this chapter we have identified attributes of a crisis leader based on common input from six leaders who successfully navigated crises in different contexts. These attributes consisted of character traits, knowledge, and abilities. Table 11.1 summarizes each attribute.

Application

1) Conduct a self-assessment to determine which leadership attributes you possess and identify opportunities to improve.
2) Identify several trusted coworkers who have had an opportunity to observe you on a regular basis. This should include at a minimum, a supervisor, someone you supervise, and lateral colleagues with whom you work. Invite them to provide feedback by rating your leadership attributes in these areas.
3) Based on Section 11.5 Developing Crisis Leadership Skills and the results of the surveys conducted in items 1 and 2, formulate professional development goals for the next five years. Practice writing these as SMART goals with a measure of effectiveness.

Example assessment: Select the degree to which each of the following statements outlined in Table 11.2 is accurate.

Table 11.2 Assessment for Crisis Leadership Attributes.

Criteria	Don't Know; Have Not Observed	Strongly Disagree	Somewhat Disagree	Somewhat Agree	Strongly Agree
Humility					
1. Demonstrates a genuine concern for others					
2. Genuinely compliments others					
3. Slow to seek attention for their own contributions					
4. Shares credit with others					
5. Emphasizes collective success over self					
6. Demonstrates self-confidence					
7. Values their own contribution					
8. Transparent and admits to mistakes					
9. Refers to "we" or "team" more than "I"					
10. Willing to take on lower-level work for the good of the team					
Integrity					
1. Actions align well with words and beliefs					
2. Makes good and moral decisions					
3. Follows through on words and promises					
4. Dependable to follow through					
5. Trustworthy					
6. Exercises wise stewardship over resources					

(Continued)

Table 11.2 (Continued)

Criteria	Don't Know; Have Not Observed	Strongly Disagree	Somewhat Disagree	Somewhat Agree	Strongly Agree
People Smarts					
1. Possesses common sense regarding others					
2. Interpersonally appropriate and aware					
3. Senses what is happening in a group					
4. Deals with others in effective ways					
5. Asks good questions and listens to others					
6. Engages in good conversations					
7. Exercises good judgment and intuition regarding the subtleties of group dynamics and impact of their words and actions					
8. Shows empathy toward others on the team					
9. Good at adjusting behavior and style to fit the context of the situation/conversation					
Moral Courage					
1. Willing to speak up for what is right despite opposition and risk to reputation/position					
2. Understands the morally right thing to do and pursues it					
3. Unafraid of political consequences					
4. Uncompromised in decision-making					
5. Acts justly even when unpopular					
Self-awareness					
1. Aware of strengths and weaknesses					

(Continued)

Table 11.2 (Continued)

Criteria	Don't Know; Have Not Observed	Strongly Disagree	Somewhat Disagree	Somewhat Agree	Strongly Agree
2. Has taken a personality survey and recalls the results					
3. Understands how their personality profile complements or conflicts with other personality profiles					
4. Recognizes when to defer to other experts due to lack of knowledge/ experience					
The Plan					
1. Familiar with the organization's emergency plans					
2. Knows their roles and responsibilities during an emergency					
3. Aware of the internal/external resources available during an emergency					
4. Understands the organization's shortfalls and limiting factors					
Situational awareness with leading indicators					
1. Knows the organization's leading indicators of success					
2. Can differentiate between leading and lagging indicators					
3. Understands how indicators affect decision-making					
Multi-disciplinary knowledge					
1. Has formal education or training beyond primary position					
2. Has experience working in other areas beyond primary education, training, or current role					
3. Has worked in a multi-discipline department					

(Continued)

Table 11.2 (Continued)

Criteria	Don't Know; Have Not Observed	Strongly Disagree	Somewhat Disagree	Somewhat Agree	Strongly Agree
Communication skills					
1. Actively listens to others, demonstrating a desire to understand their message/position					
2. Clearly articulates goals or tasks					
3. Verbal communication is clear and understandable					
4. Written communication is simple and clear					
5. Utilizes appropriate communication method based on the urgency and type of message					
Complex problem-solving					
1. Can effectively determine root causes and contributing factors to problems					
2. Applies resources to root causes instead of simply solving isolated problems					
3. Recognizes the dynamic aspect to problems					
4. Considers all stakeholders and their interests					
5. Utilizes objective (non-politicized) information					
Trust building					
1. Is open and transparent about personal matters with others					
2. Demonstrates vulnerability					
3. Demonstrates competence in their role					
4. Is respectful toward others					

(Continued)

Table 11.2 (Continued)

Criteria	Don't Know; Have Not Observed	Strongly Disagree	Somewhat Disagree	Somewhat Agree	Strongly Agree
5. Can disagree with others in a healthy, productive manner which enhances the relationship					
6. Reliable					
7. Trustworthy					
Decisiveness					
1. Makes decisions in a timely manner					
2. Does not succumb to analysis by paralysis					
3. Not afraid to be wrong					
4. Owns their decisions regardless of outcome					
5. Goal-oriented					

References

1 Lencioni, P. (2012). *The Advantage: Why Organizational Health Trumps Everything Else in Business.* San Francisco, CA: Jossey-Bass.

2 Lencioni, P. (2018). *The Ideal Team Player: How to Recognize and Cultivate the Three Essential Virtues: A Leadership Fable.* New York: Wiley.

Further Reading

1 Myers-Briggs personality test. https://www.myersbriggs.org/my-mbti-personality-type/take-the-mbti-instrument.

2 DiSC personal assessment tool. https://www.discprofile.com/what-is-disc.

3 The hiring suite pre-employment survey. http://www.thehiringsuite.com/Home.aspx?Content=HS.

4 Communication theory, Johari's window. https://www.communicationtheory.org/the-johari-window-model.

Part Five

A Safe and Secure Tomorrow

While writing this book, the world has experienced a global pandemic, military conflicts, unprecedented civil unrest, worsening natural disasters, proliferation of cyber attacks on critical infrastructure and organizations, record-setting high temperatures, droughts, excessive migration, increasing costs associated with disasters, infrastructure failures, increasing ethnic and religious persecution, spreading wildfires, and invasive species. Some disasters are confined to local jurisdictions requiring state or federal support. Other disasters are multi-national and require international cooperation. Many disasters occur within the private sector such as businesses and critical infrastructure. Our problems are complex, exacerbated by climate change, technological advances and interdependencies, globally-integrated supply chains, migration, and the growing wealth gap. The traditional approach to emergency management, a government centric and driven approach, offers a framework for crisis leaders but is lacking in the whole community approach that is necessary to create the level of resilience needed for tomorrow.

The last few chapters of this book will offer a vision for how crisis-ready leaders can prepare their organizations and communities for a safe and secure tomorrow. As you might have anticipated, this begins before the incident, but we must recognize that the delineation of incidents is blurry as there are multiple incidents occurring simultaneously and many crisis leaders are not experiencing the inter-incident pause of normal operations that once existed. Rather, we are constantly managing emergencies, disasters, and crises that are in various phases. Nonetheless, we must ready our organizations and communities for the next incident and emerging threats while the operating environment continues to evolve with all its complexities and interdependencies. We must also rise to the challenge of caring for the growing vulnerable populations within our communities and organizations that lack access, means, or function to prepare themselves for responding and recovering from the next disaster.

Prior to a crisis, a leader can prepare and build resiliency. During a crisis, a leader must navigate it well and quickly transition into recovery. These last three chapters on preparedness, building resiliency, and navigating the pathway from crisis to recovery work together and the crisis-ready leader has a role in each of these areas.

12

Preparedness

In Chapter 1, Emerging Threats and Hazards, we identified what we need to prepare for. In Chapter 3, The Challenge Ahead, we briefly considered some of the challenges facing a crisis leader such as emerging hazards, operating contexts, and vulnerable populations. Then, we made the case for whole community engagement in preparedness and building resiliency. In this chapter we will take a deeper dive into preparedness and the leader's role in ensuring that their organization and community are prepared through the preparedness cycle framework.

> **Preparedness** is defined by DHS/FEMA as "a continuous cycle of planning, organizing, training, equipping, exercising, evaluating, and taking corrective action in an effort to ensure effective coordination during incident response." [1]
>
> The US National Preparedness Goal is "A secure and resilient nation with the capabilities required across the whole community to prevent, protect against, mitigate, respond to, and recover from the threats and hazards that pose the greatest risk." [2]
>
> **National Preparedness** is defined as: The actions taken to plan, organize, equip, train, and exercise to build and sustain the capabilities necessary to prevent, protect against, mitigate the effects of, respond to, and recover from those threats that pose the greatest risk to the security of the Nation.

Learning Objectives

After reading this chapter, you will be able to:
- Describe the preparedness cycle.
- Understand the leader's role in building capabilities through the preparedness cycle.
- Identify methods for building momentum to accelerate organizational preparedness.

12.1 Preparedness Cycle

This chapter equally applies to the public and private sectors despite the differences in application and predominance of federal government structure and framework around this topic. International consensus standards such as ISO 9001 Quality Management System, NFPA 1600 Standard on Continuity, Emergency and Crisis Management, ANSI/EMAP 5-2019 Emergency Management Standard, and

Crisis-ready Leadership: Building Resilient Organizations and Communities, First Edition. Bob Campbell, PE.
© 2023 John Wiley & Sons, Inc. Published 2023 by John Wiley & Sons, Inc.

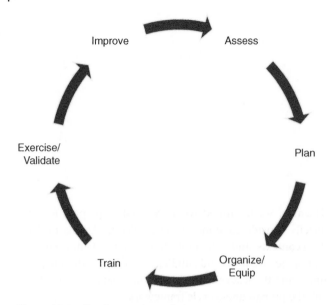

Figure 12.1 The Preparedness Cycle.

ISO 22301 Business Continuity Management System provide all organizations with frameworks for preparedness. Because the private sector owns and controls 85% of the critical infrastructure in the United States, it is essential that both sectors coordinate on preparedness activities as they engage in each of the following elements of the preparedness cycle. The interface between private operations and public safety must be well-defined, coordinated, documented, and practiced. The following subsections highlight the leader's role throughout the preparedness cycle as summarized in Figure 12.1.

12.1.1 Assessment

The preparedness cycle typically begins with a risk assessment. This typically referred to as a Hazard Identification Risk Assessment (HIRA). A HIRA identifies the numerous threats and hazards that could affect an organization or community and then estimates the likelihood and consequences of each event. A typical method defines the likely incident and estimates the likelihood of occurrence based on historical data, trends, and projection into the near future. The consequences may be estimated based on factors such as warning time, incident duration, and impact on various components of society such as the environment, people, economy, critical resources, property, continuity of government, and infrastructure. The private sector may consider impacts to people, the environment, assets, reputation, and supply chain. A more extensive analysis could be based on the Business Model Canvas components: key partners, key activities, value propositions, customer relationships, customer segments, key resources, distribution channels, revenue streams, and cost structure [3]. By accounting for both likelihood and consequences, the risk can be determined without bias to the most severe or likely incident. Decisions for resource allocation, investment, and preparedness strategies should be based on risk with the goal of lowering the risk profile through risk management. Some actions will have a favorable risk reduction across multiple hazards. Ultimately, leaders should strive to optimize risk reduction with limited resources. Leaders should ensure that a risk assessment is comprehensive to inform resourcing decisions and priorities along the remainder of the preparedness cycle. A comprehensive risk assessment addresses changing environmental conditions and emerging threats.

12.1.2 Planning

There are many types of plans (e.g., business, continuity, strategic, operations, tactical, mitigation, emergency, recovery, incident action plan, etc.). Plans often begin with an overarching objective, SWOT analysis (i.e., strengths, weaknesses, opportunities, threats), HIRA, or some assessment of the situation internally and externally. A plan seeks to frame a problem and determine how an organization will deal with that problem through a series of goals or actions. Plans anchor organizations around a common set of beliefs regarding how a situation will be handled. Risk-based planning is a process of planning that is informed by risks with priority given to greater risks. Chapter 4 explored this process in more detail.

The planning process includes the whole community of stakeholders to ensure that valid assumptions are made and diverse perspectives are included in the process. These stakeholders comprise both internal personnel to the organization and external, interested parties. A good plan describes the situation, decision points for implementing elements of the plan, roles and responsibilities, and procedures that are implemented upon reaching a set condition or criteria. Participants in the planning process learn about other organizations, their capabilities, resources, and methods for responding to incident.

Strategic plans establish a vision of success, a mission or purpose, goals, objectives, milestones, action items, and responsibilities for building future capabilities. Leaders are responsible for establishing the vision and mission for where they are taking the organization and the organization's purpose. This requires key leader and stakeholder input and buy-in. These plans are informed based on the gaps identified in the assessment, after-action reports, training, feedback, and inclusiveness of feedback from stakeholders. A solid strategic plan serves as an organization's roadmap to successful preparedness and drives toward positive outcomes from disasters. It builds capabilities and adequate capacity within those capabilities. Leaders must hold their organization accountable for tracking progress toward goals, celebrate successes, and adequately resource the plan's initiatives through decision-making.

While the strategic plan will require the most input from the leadership within an organization, leaders must be engaged in the planning process for other organizational plans, and involved in reviewing and approving plans. This may begin with a charter to establish inputs, the planning team, and expected deliverables, milestones, and schedule. Once a plan is completed and approved, leaders must ensure that the plan is disseminated to appropriate stakeholders, reviewed, trained, and tested through exercises at regular intervals; these intervals depend on the type of plan and urgency involved. Because these actions require resources and decision-making, leaders should review and approve plans at an appropriate level where the resource decision-making will occur to ensure the plan can be implemented.

Plans span and unify the five mission areas of emergency management: Prevention, Protection, Mitigation, Response, and Recovery. While organizations may have plans that are unique to one or more mission areas, the planning process spans all areas. To be effective, plans must be reviewed, maintained, and implemented as part of the preparedness cycle.

12.1.3 Organizing and Equipping

Building capabilities consists of organizing personnel, purchasing equipment, building infrastructure, establishing relationships and supply chains, and leveraging financial resources to sustain the capability. Personnel should be organized based on their role or team within the incident management system as outlined in the plans. This may require a leader to issue appointment letters to assign personnel. It is a best practice to assign the personnel by position instead of by name in order to ensure currency. Including the role in the position description can ensure continuity as managing individual appointment letters may be overwhelming and result in unfilled roles due to lagging assignments. In order to execute the responsibilities of the team or assignment, personnel must be equipped to perform the task. Equipment often consumes supplies or requires maintenance to be sustained and operational. This requires financial resources to sustain the capability throughout its lifecycle.

Leaders must make decisions and prioritize the resources required to establish the capabilities to respond based on risk. This is an area of competing interests and requires discernment over which resources will result in the greatest risk reduction. When allocating resources to improving capabilities, leaders must weigh the costs and risks associated with impacts in each mission area. For example, the cost of mitigation efforts for specific hazards should be weighed against the risk of the hazard and cost of response and recovery. Leaders are responsible for stewarding the resources entrusted to them.

12.1.4 Training

While there are many courses focused on responders, we need to recognize that all personnel within the organization require training on prevention, mitigation, response, and recovery based on the roles that they may fulfill. For example, providing cybersecurity awareness training helps prevent cyberattacks and protects information systems from unauthorized access and tampering. This is just as important as training the IT department on responding to attacks and recovering the system. In order for team members to perform competently before, during, and after an incident, the team must be trained properly.

This training takes on many forms such as classroom, knowledge-based training, hands-on, application-based training, and judgmental scenario-based training. Knowledge-based training is typically taught through web-based systems, self-paced, or in-person training. Independent study courses typically fall into this category. Hazmat Awareness is an example of knowledge-based training that teaches participants how to recognize hazardous materials through placards. Hands-on, application-based training involves utilizing equipment to become proficient in a given task. For example, decontamination training typically involves setting up a decontamination system and using it to perform the task of removing contaminants from people or objectives. Scenario-based training involves immersion into a situation where participants must make decisions to resolve some problem. This requires application of knowledge, skills, and abilities to make decisions and formulate actions to resolve the situation. In addition to these levels of training, training may be delivered informally, formally, or in a multidisciplinary environment.

While it easier to implement knowledge-based training, leaders must assess the level of training needed for a person to adequately perform their job based on their roles and responsibilities. After reviewing several Chemical Safety Board investigations into various incidents, it is apparent that many of the first responders involved in major incidents were not trained to the level necessary for the role they were playing in the response. In assessing the appropriateness of training, leaders should not only review the course description but also the course learning objectives to evaluate whether the training outcomes align with performance expectations as outlined in the position description. Leaders should also consider the role of credentialling especially as most crises will result in multiagency response actions. Credentialling enables leaders to understand and accept (in advance) the common competencies expected from multiagency responders. Without credentialling, leaders open themselves and their organization to the risk of engaging unqualified personnel in risky operations. Credentialling establishes a common standard for competencies and ensures that those engaged in response are qualified to do so. Finally, leaders should monitor training currency among their team to ensure that everyone is trained properly to fulfill their roles and responsibilities before, during, and after an incident.

My company, Alliance Solutions Group, Inc., has worked with major metropolitan regions in conducting training needs assessments to help these communities better define their training needs across multiple disciplines given the spectrum of threats and hazards, and current training. The key step in the process was establishing criteria for determining training gaps. We have also served as a FEMA training partner for over five years designing, developing, and delivering training that addresses emerging needs throughout the United States. FEMA's standards for instructional system design are exacting, and training products undergo multiple reviews to ensure the highest quality. Course objectives, topics, lessons, and duration of each are mapped across their respective levels of knowledge, skills, and attitudes to determine the appropriate type of course (i.e., awareness, intermediate, management, advanced, etc.) and ensure alignment with the needs determined during the needs analysis. Often, we find that the missing link is leadership or stakeholder awareness of the required level of training. We help ensure that the right level of training is implemented for our customer which may include our nation's first responders, emergency managers, economic development organizations, hospitals, etc. As we have trained crisis leaders and responders around the world, we have identified several important considerations that lead to success: understand the culture and the operational context, know the capabilities and limitations of the participants, strike the right balance among knowledge, hands-on, and scenario-based training based on their needs and objectives, and make it realistic by incorporating innovative methods such as virtual/mixed/augmented reality, emulation, and dynamic responses. To prepare for the future, leaders should train in multidiscipline, multiagency environments to learn from different perspectives and expand their operating context. This enhances the participants' ability to interact in these operating environments while sharing best practices, techniques, technologies, and systems that may also improve organizational capabilities through adoption and integration.

12.1.5   Exercise/Testing

After developing a capability through planning, resourcing, and training, it is important to measure the effectiveness of these actions in reducing the community or organization's risk profile established in the first step of assessing the risks. This can be accomplished through a series of testing or exercising to verify that the system works and validate the effectiveness of the capability.

Exercises can be discussion-based or operations-based. Discussion-based exercises include seminars, games, and tabletop exercises. Operations-based exercises include drills, functional, and full-scale exercises. An exercise series comprised of some or all of these types of exercises is a good way to test various aspects of the capability in appropriate contexts through escalating steps. For example, an organization may test its plans through a seminar or tabletop exercise. A health-care coalition may test their surge capacity plan through a game or advanced tabletop exercise. A fire department may test their hazmat response capabilities in a drill. An EOC may test their coordination and communication systems with external parties through a functional exercise. A company may conduct an annual full-scale exercise to validate catastrophic emergency procedures.

Objective-based exercises seek to verify or validate specific capabilities based on objective criteria that are incorporated into the exercise evaluation process. A realistic, risk-based scenario is used as a context within which to test the capabilities. Exercises should mature into multidiscipline, multi-jurisdiction exercises which include all of the stakeholders that may be involved in incident response and recovery. Incorporating the transition from pre-incident to response and then recovery can serve as a valuable objective to ensure the team knows how to make these transitions.

Leaders need to know that their organization is prepared to deal with the predominant risks identified in their risk profile. Exercises offer this opportunity and enable the team to gain experience working in a multidisciplinary, multi-jurisdiction environment. The exercise should result in opportunities to improve.

Alliance Solutions Group, Inc. has found the following keys to successful exercises based on designing, developing, and conducting 2,000+ exercises in 48 states and 17 countries:

- Establish roles for members of the planning team early in the process.
- Identify objectives around the aspect of the system that needs to be tested. This will determine the type of exercise needed and the participants; some organizations decide on the type of exercise, the scenario, the participants, and then the objectives – this is backwards.
- Assign a lead developer for the exercise-development team with subject-matter experts to support development; do not attempt to design and develop an exercise by committee otherwise it will be disjointed.
- Focus on controlling and evaluating the aspects of the system that align with the objectives and criteria in the evaluation guide; avoid rabbit trails, "what if" scenarios, and "calling audibles."
- Ensure realism and safety through simulation, emulation, and proper role playing; never compromise the safety of participants for hazards, props, and special effects that are unnecessary for task performance.
- Assign an expert as the controller/facilitator so that when the exercise goes off script, the controller/facilitator can detect it and appropriately redirect through questions, injects, or other natural methods.
- Train the evaluators on how to use the evaluation guide and their responsibilities regarding observations, notes, and after-action reporting.
- Gather participant feedback and evaluator observations immediately after the exercise. Sometimes it's a long day, but capturing this while it is fresh is critical considering the cost of the exercise and the risk of losing this information.

12.1.6 Improvement

If the exercise was the test of the system, then it is necessary to report the results of the test in terms of system effectiveness. This is typically accomplished through an after-action report/improvement plan. The evaluation begins with identifying deficiencies in the system. There are numerous ways to organize these evaluations, but I recommend using the method that best aligns with driving improvements in your organization or community. The most common is the Homeland Security Exercise Evaluation Program (HSEEP) After Action Report/Improvement Plan (AAR/IP) template. This approach organizes information based on the capabilities, activities, and tasks performed. Another approach is to evaluate the topical elements that comprise the capability (e.g., plans, organizational structure, training, personnel, facilities/infrastructure, leadership, equipment). Regardless of the reporting structure used, leaders should be interested in the gaps so that they can take action to improve their organization's preparedness. Having collaborated with many organizations in writing AAR/IPs, it is not uncommon to encounter someone that wants to soften the problem or gap identified to avoid making someone look bad for political reasons. This is where an independent, third-party can be helpful in being objective as they document observations, the review process, and rationale behind any change to the report. Leaders would be better served to insist on receiving the draft AAR/IP from the third-party, and then coordinating with the other stakeholders for the review process.

The AAR/IP should inform a leader's strategic planning and investment priorities, as they continue to reduce and manage risks. The preparedness cycle continues back to the assessment and so on.

But exercises and audits are not the only input that should drive improvement. Feedback and outcome analysis from training, ideas from those within the organization, discussions resulting from the planning process, user feedback on equipment, risk assessments, etc. should also inform leaders on how they can improve their capabilities. An engaged leader becomes more aware of the organization's capabilities, limitations, and vulnerabilities by participating in these processes. While many leaders do not have the time to fully participate, leaders can stay informed through updates from key staff members, listen/sit-in on some part of meetings, train based on skills needed (closing gaps), and participate in exercises. Leaders should monitor organizational improvement through the strategic planning and oversight process. Elected officials also have a role in the Policy Group which is an extension of the Emergency Operations Center; elected officials should engage with emergency managers prior to an incident to become familiar with policies, procedures, and practices through training and exercise participation. Leaders should invest in leadership development within their organization by allowing rising leaders to participate in leadership roles during training and exercise events.

12.2 Preparedness Cycle Implementation

Following the preparedness cycle ensures continuous improvement and enhances the organization's preparedness through a systematic process. Organizations sometimes encounter internal resistance to publishing "observations" and feedback that could drive improvements. This requires leaders with humility that welcome feedback and encourage observations to drive improvements. Most organizations encounter challenges to implementation when they engage external stakeholders and partners in the process. Building a committed coalition of partners who are willing to work toward common capabilities can be cost effective while spreading out the demands among many organizations and teams. This requires leaders to engage in outreach to build external relationships and make introductions so that their team can engage with leadership buy-in. Regional risk assessments, planning, training, exercises, and mutual-aid agreements can pay dividends by drawing from more resources, coordinating capabilities and targets, sharing risks and costs, and building mutual trust. Regional approaches have been successful in many metropolitan areas (i.e., multijurisdictional planning districts) but have also been used in the health-care sector (i.e., health-care coalitions).

In practice, there are short-circuiting cycles within the preparedness cycle that are occurring simultaneously. For example, the planning process may highlight an improvement without waiting for the exercise. The same is true with training, equipping, and organizing. The faster that an organization can cycle through the improvement process, the more rapidly they improve their capabilities and preparedness. A good leader facilitates these cycles.

In summary, a leader's role is to establish the vision for the organization and community. Strategic planning is a crucial step in this process that requires leadership participation, buy-in and follow-up. It sets the cycle in motion. By establishing a charter and expectations for their team to engage in planning, the leader drives the process and creates priorities and direction. As the planning process identifies resourcing requirements, the leader communicates support by allocating resources based on organizational risk reduction. Throughout the cycle, the leader must welcome and encourage their team to identify gaps, provide feedback, and conduct honest assessments. This fosters an environment of openness ripe for continuous improvement and builds momentum to accelerate the preparedness cycle. This results in a highly capable organization that is prepared. However, if the leader squashes feedback and suppresses "bad news," the opportunity to improve never surfaces and the cycle stops. By monitoring and measuring organization performance along the preparedness cycle, leaders demonstrate interest and priority for establishing capabilities that result in a prepared organization.

Summary of Key Points

- A leader's role in preparedness is to stay informed and engaged throughout the preparedness cycle.
- A solid strategic plan serves as a leader's roadmap to successful preparedness and determines outcomes from disasters.
- Planning, resourcing, training, and exercise decisions should be informed based on a comprehensive Hazard Identification Risk Assessment.
- Resource allocation decisions should be prioritized based on risks to optimize risk reduction.
- Form multidiscipline, multiagency coalitions to coordinate and collaborate on grant opportunities, funding allocation, and whole community preparedness.
- Job responsibilities (knowledge, skills, abilities) determine training needs; learning objectives should align with performance expectations.
- Proper exercise design, development, and conduct are necessary to verify and validate capabilities. This is essential to inform decision makers on gaps.
- Close gaps identified after action reports, training feedback, planning processes, and engagement from stakeholders.
- Facilitate the organization's progression through the preparedness cycle and build momentum to continuously improve.

Keys to Resilience

- A prepared community has the potential to be a resilient community.
- The adage, "an ounce of prevention is worth a pound of cure" should guide risk-based decisions to allocate funds for prevention, protection, and mitigation to save on response and recovery which can be more costly.
- A whole community approach to preparedness ensures that vulnerable populations are integrated into planning and resource allocation and the private sector needs are identified to sustain, stabilize, and restore community lifelines.

Application

1) To what extent does your organization follow the preparedness cycle? Give examples of how accomplishing initiatives in each element of the preparedness cycle over the last year has enhanced your capabilities.
2) How are leaders within your organization engaged in each element of the preparedness cycle?
3) Identify the benefits of leadership participation in each element. How can this participation enhance their leadership during a crisis?
4) What are your organization's strategic planning goals? How does leadership monitor and measure performance and effectiveness of these goals throughout the year?
5) What is your strategy for allocating resources to reduce the most risk?

References

1 Federal Emergency Management Agency. Plan and prepare for disasters (archived content). https://www.dhs.gov/plan-and-prepare-disasters (accessed January 8, 2022).
2 Federal Emergency Management Agency. National preparedness goal. https://www.fema.gov/emergency-managers/national-preparedness/goal (accessed May 27, 2022).
3 Osterwalder, A. and Pigneur, Y. (2010). *Business Model Generation*. New York: Wiley.

Further Reading

1 FEMA's preparedness toolkit index page. https://preptoolkit.fema.gov.
2 Economic Development Administration's Disaster Recovery index page. https://eda.gov/disaster-recovery.

13

Building Resilience

Chapter 12 outlined a leader's role in preparedness – a process of building capabilities across the mission areas of emergency management based on risks. Building resilience complements preparedness by enhancing the system in ways that minimize impact and ensure rapid recovery from stressors. This chapter will define resilience and present several approaches to creating models for examining resilience within a system. Applying components from several of these models, this chapter outlines a systematic approach for examining resilience in a system and developing strategies to enhance resilience. Given our interdependent environment, crisis-ready leaders are engaged with the whole community of stakeholders to anticipate adverse impacts on the system and develop strategies to enhance resilience for the good of their organization and community.

Learning Objectives

- Define resilience.
- Identify how preparedness actions across the mission areas of emergency management can contribute to resilience.
- Identify indicators and determinants of resilience.
- Utilize a resilience model to examine a system and develop strategies for enhancing resilience.
- Identify the role of leaders in creating a resilient community or organization.

Resilience Definition
   Meriam-Webster Dictionary defines **Resilience** as:

1) the capability of a strained body to recover its size and shape after deformation caused especially by compressive stress;
2) an ability to recover from or adjust easily to misfortune or change.

Community Resilience is defined by National Institute of Standards and Technology (NIST) as "the ability of a community to prepare for anticipated natural hazards, adapt to changing conditions, and withstand and recover rapidly from disruptions." [1]

Crisis-ready Leadership: Building Resilient Organizations and Communities, First Edition. Bob Campbell, PE.
© 2023 John Wiley & Sons, Inc. Published 2023 by John Wiley & Sons, Inc.

13.1 A Resilience Model

Enhancing resilience can reduce the consequences associated with a hazard and thereby reduce the risk. Resilience has two main elements: a structure or boundary, and structural properties which can rebound or recover from stress. This implies that there are stressors in the environment which can strain the structure within the boundaries and cause stress. The structure or body consists of interconnected, interdependent systems or members. As stressors affect some part of the structure, other parts may also be impacted. The goal of resilience is to analyze the various stressors in the environment and their impact on the structure to include its members and systems, and then incorporate components that enable resilience (i.e., the ability to recover). Some may take this a step further by designing resilient components which not only recover but recover faster and better thereby improving upon the original state of the structure.

While an entire book could be written on the topic of resilience, we are going to broadly explore several types of "structures" such as a community, corporation, and to some extent an individual, and address how to incorporate resilience into these structures to enable quick recovery from a crisis.

The 6 R's of Resilience outline several attributes of resilient systems that can be used to formulate resilience measures or evaluate the extent of a system's resilience.

Ready: How well is the system prepared for the hazards and threats outlined in its risk profile? How prepared is the organization to take advantage of opportunities in a timely manner?

Redundant: Are redundancies built into essential components of the system? e.g., redundant sources of supply, methods for product/service delivery, key staffing, financing and banking, revenue/customer diversification, etc.

Repairable: How easily and quickly can an essential component of the system be repaired if it is damaged? Are the supplies, resources, and procedures in place to return the damaged component to an operable state within the required timeframe?

Resistant: How resistant to damage are the essential components of the system? e.g., are structures built to a standard that can resist or prevent damage from hazards such as flooding, wind, hail, blasts, etc.?

Robust: How strong and persistent are the system and its components when facing disruptions? Is the system mature, defined, managed, scalable, replicable, etc.?

Reserves: Have leaders adequately established and maintained reserves such as sufficient cash to maintain operations during cash flow interruptions, liquid assets that can be converted to cash, excess personnel capacity to deal with staffing shortages, adequate inventory reserves, or additional space to accommodate physical distancing or take advantage of process modifications?

13.2 Resilience Indicators

FEMA has recently adopted and incorporated the University of South Carolina Hazards & Vulnerability Research Institute's Baseline Resilience Indicators for Communities (BRIC) [2] into the National Risk Index. While there are multiple resilience indicators and models, we will refer to this model for the community context as it pertains to crisis leadership and decision-making. The BRIC index utilizes 49 variables across six broad capitals as input indicators for computing resilience to natural hazards. While this model does not include technological or man-made hazards, there is sufficient overlap to establish validity. These capitals and some sample variables include:

Human Well-being/Cultural/Social – physical attributes of populations, values, and belief systems (educational attainment equality, preretirement age, personal transportation access, communication capacity, English language competency, non-special needs populations, health insurance, mental-health support, food security, access to physicians).

Economic/Financial – economic assets and livelihoods (homeownership, employment rate, racial/ethnic income inequality, nondependence on primary/tourism sector employment, gender income inequality, business size, large retail with regional/national distribution, federal employment).

Infrastructure/Built Environment/Housing – buildings and infrastructure (sturdier housing types, temporary housing availability, medical care capacity, evacuation routes, housing stock construction quality, temporary shelter availability, school restoration potential, industrial resupply potential, high-speed internet infrastructure).

Institutional/Governance – access to resources and the power to influence their distribution (mitigation spending, flood insurance coverage, governance performance regimes, jurisdictional fragmentation, disaster aid experience, local disaster training, population stability, nuclear accident planning, crop insurance coverage).

Community capacity – social networks and connectivity among individuals and groups (volunteerism, religious affiliation, attachment to place, political engagement, citizen disaster training, civic organizations).

Environmental/Natural – natural resource base and environmental conditions (local food supplies, natural flood buffers, energy use, perviousness, water stress).

Another approach is to utilize the Community Lifeline categories and components as discussed in Chapter 7. Whichever approach is decided, leaders can measure, evaluate, and make decisions that improve upon their community resilience.

A similar approach can be implemented for business resilience. The leader's role is to create a culture of resilience by understanding their risks, risk tolerance, and rate that the risk profile is changing. This drives business leaders to evaluate their capabilities compared with their vulnerabilities and then design resilience into their systems. Taking a capital approach to determining resilience indicators for businesses leads to the following:

Human Capital – employees, vendors, access to qualified labor; recruiting, hiring, onboarding systems, feedback/evaluation systems; indicators may include utilization, turnover, productivity, customer satisfaction, and compensation equity.

Infrastructure – facility, equipment, supplies, internet, transportation, cyber-infrastructure; indicators may include adequacy of built environment for hazards, internet bandwidth, supply chain capacity and redundancy, power reliability.

Financial Capital – cash reserves, customer diversification, backlog, liquidity, assets, insurance; indicators may include cash reserve/current liabilities, percentage revenue from largest customer, backlog/annual revenue, current assets/current liabilities, insurance coverage/assets.

Intellectual Capital – intellectual property, knowledge management and preservation, information systems.

Governance/Systems – management systems that include policies, procedures, assignment of responsibilities, and associated plans such as business continuity.

A more elaborate business resilience model and assessment was developed by The Ohio State University, the Supply Chain Resilience Assessment and Management (SCRAM™) [3]. This assessment tool accounts for 14 capability factors and 7 vulnerability factors to rate an organization's resilience.

13.3 Traditional Response and Recovery Model vs. Simultaneous Recovery

Under a traditional emergency management model, the five mission areas of prevention, protection, mitigation, response, and recovery are sequential and compartmentalized. In reality, there is some overlap, especially between response and recovery operations which can occur concurrently in different domains (e.g., the COVID-19 pandemic has been a series of ongoing prevention, mitigation, response, and recovery actions that have often overlapped). During a crisis, response actions are taken to stabilize the incident, and then recovery actions are taken to restore the damage from the incident and reset conditions to a normal steady state.

For example, more than 46 million residences in 70,000 communities are at risk of wildland urban interface fires due to a growing wildland urban interface [4]. During a wildland fire, families may be displaced from their homes and provided temporary housing until they can rebuild or relocate. The recovery process should begin as soon as they are displaced, while the response actions may continue for weeks to extinguish and control the wildfire.

> The **Wildland Urban Interface** (WUI) is the zone of transition between unoccupied land and human development. It is the line, area, or zone where structures and other human development meet or intermingle with undeveloped wildland or vegetative fuels [5].

Often, decisions made during the response will impact the recovery since recovery occurs with the response. These decisions could ultimately impact the extent of the damage and duration of recovery. These are crisis decisions. Using our wildland fire example, decisions on where to deploy resources to control the fire and minimize damage to homes can affect the number of displaced residents and damage to property. Early access to mutual aid or federal resources can preempt expansion of the fire. Decisions on insurance valuation and companies with fast claim processing times can accelerate recovery for the residents. A business continuity of operations plan may enable a business to pivot to an alternate facility and recover files electronically from a back-up server. However, if key IT personnel are affected by the evacuation from their home, they may be delayed in activating the continuity plan for their employer. This is an example of interdependency.

The crisis leader's goal is to make good decisions early to steer the response into recovery while minimizing the impact. Because a crisis like a wildfire could quickly overwhelm resources, a leader's role begins well before the event. For example, the leader has a prevention and mitigation responsibility to advocate for measures that would prevent the crisis such as restrictions on campfires, personnel access limitations to dry areas, controlled burning and harvesting forests where applicable, creating no-growth zones to contain a possible fire, lobbying against zoning practices that would allow residents to build deeper and closer into fire-prone areas, lobbying for building code changes to protect residents from likely hazards, and participating in exercises that demonstrate sufficient capabilities and response timing that are needed to quickly contain and stabilize the incident.

13.4 A Systematic Approach to Resilience

Different hazards will present different stressors and subsequently impact the system in different ways. The following steps outline a process for building resilience. This process overlaps and parallels with the preparedness cycle. One way to think of the relationship between preparedness and resilience is that resilience can be an aspiring goal within the preparedness cycle. An organization can be prepared without achieving resilience, so the resilience goal entails considering how each step of preparedness builds resilience. Another way to think of the relationship is that a culture of preparedness integrates resilience in order to reduce the consequences and risk for the hazard. Resilience reduces risk.

1) Define the system (e.g., community, business, school, economy, market, etc.).
2) Identify local, relevant hazards and assess the risk. These are the stressors that will be applied to your system.
3) Identify the components of the system that are impacted by the hazards.
4) Identify interdependent components and the extent of impact.
5) Aggregate the impact on all components based on risk of each hazard.
6) Determine vulnerabilities within the system.
7) Develop strategies to strengthen components to enhance resilience.
8) Evaluate and prioritize resilience measures that increase system-wide resilience.
9) Implement actions to increase resilience.

13.4.1 Define the System

For the purpose of this discussion, we will refer to a system as an organization (e.g., business, school, etc.), a community, or an extensive relationship among elements of the system (e.g., market, sector, or industry). Throughout this chapter, we are going to explore examples of a community and a business, but this same process could be applied to a smaller system with well-defined boundaries. To define the system as a community, it is best to start with either the jurisdictional boundaries or census tract depending on how data sets are organized. Resources such as FEMA's National Risk Index [6] can be helpful in aggregating and analyzing risk, social vulnerability, and resilience data sets by political jurisdiction.

A business may choose to define its boundaries for the entire company or individual facilities at specific locations given unique operations which align within the control of a corporation or plant manager.

Other applicable boundaries might be set around an economic sector such as manufacturing base, or an industry such as the service industry. The North American Industry Standard Codes (NAICS) may be helpful in slicing economic activities and drivers that influence resilience indicators.

13.4.2 Identify Local, Relevant Hazards and Assess the Risk

This step, also referred to as the Hazard Identification Risk Assessment, was discussed in Chapter 12 so we will not repeat it here other than to overview the two parts of this process. First, identify hazards (man-made, natural, and technological) that could impact your system. The FEMA National Risk Index tool contains a comprehensive list of natural hazards along with associated expected annual loss data which includes exposure, frequency, and historic loss-ratio information for each hazard. Second, there are multiple methods for assessing risk. The most traditional risk assessment method is to estimate the likelihood and severity of a defined incident (e.g., the likely winter storm that would shut down major roadways; category 3 hurricane, EF2 tornado, ransomware attack on the city's server). Without a clear definition of the incident, the risk assessment process will lead to some erroneous and often exaggerated conclusions. Standardize the approach to defining severity in terms of consequences (e.g., 1 to 5 scale on various impacts, warning time, and duration of the hazard). Compute a risk rating for comparison. Taking a standardized approach to all hazards will lead to comparable results (i.e., relative risk ratings). Alternatively, utilize the method outlined in the National Risk Index which multiplies the expected annual loss by social vulnerability [7], and divides by a community resilience index [2]. This approach will work for natural hazards, but the National Risk Index does not include technological or man-made hazards. Regardless of the method utilized, the goal of a risk assessment is to prioritize hazards based on risk in order to prioritize allocation of resources to optimize preparedness.

13.4.3 Identify the Components of the System That Are Impacted by the Hazards

Community Lifelines are a great starting point for understanding the components that are necessary for a community. Table 13.1 expands beyond the community lifelines and incorporates other components as outlined earlier, and crosswalks each component with a community, corporate, and individual perspective on resilience by listing various elements of each component within the respective perspective. As we explore resilience components, consider the strength, interdependence, and vulnerability of each element. Listed under each element are examples (in italics)

Table 13.1 Components and indicators of resilience.

Component	Community	Corporate	Individual
Governance	Governance, authorities, rule of law, laws and regulations, local government	Governance, by-laws, organizational documents	Moral and ethical conduct, rights and liberties
	Continuity of government plans	*Continuity of operations and succession plans*	*Family structure, wills and power of attorney, exercise of rights and privileges*
Safety and security	Public safety and security, law enforcement, judicial system	Physical, cyber, and personnel security	Personal safety and security
	Crime rates, law enforcement capacity	*Facility and cyber access controls, antivirus, HR policies*	*Home security system, firewall on Wi-Fi*
Economic	Financial stability	Revenue and profitability	Employment, insurance
	Economic diversity, insurance for publicly owned, critical facilities; regional GDP, employment, new permits	*Customer diversity and balance, backlog, return on assets (e.g., intellectual property, plant, equipment, inventory, investments), Liquidity*	*Net worth (positive), Liquidity, Income, Credit rating*

(Continued)

Table 13.1 (Continued)

Component	Community	Corporate	Individual
Resource accessibility	Access to diverse and abundant resources to support the business and population base	Access to human, technological, and transportation resources needed to add value, distribute, and sell a product or service	Access to housing, food, water, education, transportation, health care, communication, energy, social services, safety and security
	Stable community lifelines		
Supply chains	Banking, transportation systems, ports, commercial zoning	Supply chain and logistics, inventory	Mail service, stores, sources of supply
	National and community banks, lending capacity, reliable transportation infrastructure, permitting	*Redundant supply chains, reliable logistics networks, access to transportation hubs, sufficient inventory*	*Access to mail service, retail stores*
Infrastructure	Infrastructure: utilities, zoning, transportation systems, property rights, schools, hospitals/health care, communication systems	Buildings, utilities, communication systems, intellectual property	Home (house, apartment, living facility)
	Access to ports, airports, reliable roads, railroads, pipelines; education system capacity; available real estate, zoning plans; health-care capacity and capabilities; communication systems (internet speed and cell coverage)	*Internet speed, cell coverage, access to transportation; commercial real-estate capacity, utility service reliability*	*Home equity, insurance,* *alternate housing plan*
Social and cultural capital	Human capital, access to skilled and knowledgeable workforce, access to education and housing	Human resources	Health, education, assets and income, transportation, psychological
	Demographics (age, income, education, linguistic ability, mobility)	*Labor capacity*	*Health and wellness (BMI), net worth, mental wellness, mobility*
Natural resources	Environmental protection and management in place, parks and outdoor recreation, weather	Environmental impact (e.g., environmental aspects and impacts are understood)	Environmental impact is managed and minimal
	Adequate stormwater management, protection, solid-waste management, wastewater systems, safe and abundant drinking water, free of environmental hazards, availability of outdoor recreation, climate	*Impact on environment defined and managed*	*Utilization of environment, environmental value*

of indicators that can be used to assess the strength of that element. These are just examples to provide some context so there may be other elements and indicators worth considering. What other elements would you include under each component of resilience? Any other indicators that you would use? How would you measure each indicator and determine its effectiveness?

There are several determining factors related to each component: access, capacity, reliability, insurance, plans, and adequacy. If a hazard has an impact on any of these determining factors, a delay or disruption could affect the component or the system. In the next section we will take a closer look at how a hazard's impact can affect another component based on interdependencies.

13.4.4 Identify the Interdependent Components and the Extent of Impact

Incidents can impact individuals, businesses, and communities in unique ways; however, these impacts can have secondary impacts on each as well. Let us explore a few examples to illustrate the interdependencies and cascading impact. Table 13.2 outlines a series of hazards and provides

Table 13.2 Examples of interdependent and subsequent impacts from hazards.

Hazard	Impact 1	Impact 2	Impact 3	Impact 4	Impact 5
Cyber-attack on pipeline	Fuel distribution tracking disrupted	Unable to account for distribution to vendors and collect revenue for fuel delivery	Shut down pipeline as both precautionary measure for other potential attacks and to restore tracking and invoicing	People can't buy gas	Worker cannot go to work
Pandemic	25% workforce ill or at home	Hospitals at capacity, limited care	Surgeries cancelled	Lost revenue; furloughs, turnover	Cost of health care increases
		Businesses lack workers	Production slows, output and revenue decreases	Supply chain disrupted, inventory variations	Consumers lack products
Hurricane	Trees knock out power lines, homes; flooding	Community loses power, HVAC, refrigeration	Heat stress, food spoils, crime increases, businesses shutter	Food shortage, hospitalization, production slows	Business and consumers lack supplies, food deliveries
		Transportation disrupted	Emergency vehicles cannot respond, residents cannot evacuate or go to work	Shortage of workers, supply chain disruption	
Civil unrest	Safety and security decrease, governance threatened	Injuries, deaths, injustice, property damage	Businesses shutter/leave community, insurance increases	Food deserts, job loss, business costs increase	Consumer prices increase

several examples of primary through quintenary impacts based on the principle of interdependency. In other examples there may be multiple, cascading impacts based on the factors mentioned.

> **Cascading impacts** are those impacts that result from a previous sequential impact (e.g., secondary, tertiary, etc. impacts) within a system.

I will explain the first example further; it is based somewhat on the Colonial Pipeline cyberattack based on general information available at the time of this writing. Criminals conducted a cyberattack on the pipeline's accounting system which prevented it from being able to track the distribution of fuel from the pipeline to vendors who transport fuel to gas stations and other re-sellers. Since they were unable to track offloaded fuel to vendors, they were not able to bill these vendors for the fuel that they withdrew from the pipeline. The company shut down the pipeline as a precautionary measure to ensure accurate tracking and billing of fuel but also to ensure there was no other malicious software attacks on other components of their operation. As a major fuel supplier, the disruption prevented gas trucks and rail transportation of the fuel to resellers causing gas stations to run out of gas supply. This incident transpired over about one week. As fuel became scarce in some regions, workers were unable to refuel their cars and go to work, and so on. This example illustrates the principle of interdependency where one disruption affects some other part(s) of the system which in turn has another adverse impact. As these effects ripple through the system, the impact has a tendency to grow rather than subside. Take a look at the other examples.

Another framework for evaluating interdependencies is to evaluate the supply, production, and demand around the system. Realize that the supply for one system could be the demand for another system; this is the concept of a supply chain. Let us review the interdependencies related to a business impacted by a hurricane and flooding in Table 13.3. In this table, the supply disruption caused by the hurricane and flooding incident has a downstream impact on production which in turn causes lost services, inability to deliver products, and ultimately results in decreased revenue.

Table 13.3 Example of supply chain impact from a hurricane and flooding incident.

Supply	Production	Demand
Raw materials: Transportation disruption	Shuts down production line	
Utilities: Power outage	Cannot accept and process orders, equipment is inoperable	Products do not ship, services not provided, lost revenue
Workers: Personnel at home recovering from damage to homes	Labor shortage	

Now, let us consider the impact of the pandemic on the same business in Table 13.4. The first line provides an example of shifting from one channel of delivery to another which has the potential to increase supply of services to end-customers at a lower cost. The second example shows how a shift in market conditions could result in increased costs.

Table 13.4 Examples of supply chain impacts from a pandemic.

Supply	Production	Demand
Workers: Personnel at home due to isolation, quarantine, or gathering restrictions	Inability to deliver in-person services; increased dependency on cyber infrastructure	Shift in market expectations, eliminating need for in-person services, increase in virtual services
Travel: Limited transportation options such as airline or rental cars	Cost increases, limited flexibility	Some customer needs are unmet; increased risk for clients; lower profit margins as costs increase

13.5 Aggregate the Impact on All Components Based on Risk of Each Hazard

As we examine the impact of each hazard on the system (i.e., in this case a business), we can quantify the impact in terms of lost revenue, but this could also be measured based on lost workdays, lost lives, injuries, distribution channels, key assets, or other impacts. Table 13.5 crosswalks the impact of each hazard on each component and weights the impact based on a notional risk rating; this risk rating would normally be derived from the risk assessment method chosen by the evaluator. A rank-ordered list can provide some prioritization for further analyzing vulnerable components within the system. This is where leaders recognize the need to prevent or mitigate the impact. In our example below, economic, resource accessibility, and supply chains are the top three impacted components based on the limited hazard evaluated.

Table 13.5 Component impact analysis.

Components	Hazard 1: Hurricane Risk: 75	Hazard 2: Pandemic Risk: 60	Risk-weighted impact on component; rank-ordered from highest (1) to lowest (8)
Governance	No impact	Low impact	7
Safety and security	Negligible	High impact	5
Economic	High impact	High impact	1
Resource accessibility	High impact	High impact	2
Supply chains	High impact	Med-High impact	3
Infrastructure	Low impact	No impact	6
Social and cultural capital	Med impact	High impact	4
Natural resources	No impact	No impact	8

13.6 Determine Vulnerabilities within the System

Vulnerabilities are weaknesses in the system that cause strain when the stressors from hazards impact the system. Some components of the system are more vulnerable than others depending on the hazard. Using the same list of components, reordered based on risk-weighted impact, Table 13.6 identifies example vulnerabilities associated with each component with the same example business and hurricane and pandemic hazards. These vulnerabilities can be identified by examining the root causes of the impact to the component. Within each component, there are many potential elements as we outlined earlier, but is worth highlighting diversity within the human capital component, specifically the need to account for and address vulnerable populations. This was defined and explained in previous chapters, so I will limit the discussion to leadership responsibility in creating resilient organizations and communities. Leaders are responsible for leading people within their organization or community, and the population within these groups has diverse needs. Additionally, a significant amount of resources may be expended during an incident without adequate planning for those with functional and access needs. Resilient organizations and communities account for this in the built environment through universal design and understanding the needs of their population. Strategies for addressing vulnerable populations are incorporated into plans and built into communications, transportation, information systems, resource management and projections, and emergency services. This is just one example of diversity within a component.

Strategies can be developed to strengthen these components and enhance resilience. Leaders should engage their team to ensure that they are identifying root causes as well as vulnerabilities.

Table 13.6 Vulnerabilities associated with each component.

Components	Vulnerabilities
Economic	Single source of revenue based on production and sale of products; limited backlog of orders that can be filled based on current inventory; insurance policy does not cover lost revenues.
Resource accessibility	Workers unable to travel to work when roads are closed; workers vulnerable to contracting disease, potential exposures, caring for sick family
Supply chains	Single source of supply
Social and cultural capital	Segment of workers that have comorbidities which could result in more severe illness from the virus
Safety and security	Lack of cyber-security on network; workers more likely to become injured during recovery operations, or ill from home-transmission
Infrastructure	Dependency on delivering raw products over roadways; inadequate bandwidth for working from home; network is not secure
Governance	Policies do not address working from home, leave during disasters, implementation of pandemic mitigation and prevention measures
Natural Resources	N/A

13.7 Develop Strategies to Strengthen Components to Enhance Resilience

To enhance resilience, we can prevent the impact of the hazard by removing the pathway for impact, mitigate the extent of the impact by strengthening the weaker components of the system in ways that increase capacity or capabilities, and enable swifter, more effective response and recovery. Because many hazards can have similar impacts and effects, addressing vulnerabilities can increase resilience across a multitude of hazards. Table 13.7 continues to build on our example with strategies to address each vulnerability. By ensuring that a root-cause analysis is conducted, leaders can lead their team in formulating strategies that address both the vulnerability and root cause for lasting impact.

13.7.1 Evaluate and Prioritize Resilience Measures That Increase System-wide Resilience

At this point, we have identified a series of resilience initiatives that can enhance resilience for a business. But due to limited resources, an entity will not be able to implement every measure. There is a point of diminishing return where further increases in resilience measures can negatively impact the business due to the untenable costs (i.e., another business risk which could cause the very bankruptcy that we are trying to prevent). It is impossible to eliminate all risks and vulnerabilities. Therefore, it is important to evaluate and re-prioritize the resilience initiatives based on their return on investment. Depending on the risk methodology utilized, we may need to evaluate how the resilience initiative changes the risk based on its inputs. For example, if you use a risk method like the one described above based on likelihood and consequences (which are comprised of various impacts, duration, and warning time), update these factors based on the implementation of the resilience initiative to determine how much risk is reduced. If you are using the FEMA expected annual loss input from the National Resilience Index, you could estimate an expected percent reduction based on the initiative. Be consistent in your approach so that the results are comparable. The example in Table 13.8 will apply notional figures based on reduction in risk rating once the measure is implemented (e.g., increasing inventory of raw materials for the expected duration of the disruption due to a hurricane might eliminate the risk associated with the supply chain component but not for the winter storm hazard). Here is an example to illustrate this evaluation and re-prioritization of resilience initiatives. The risk rating associated with each hazard is listed below each respective hazard. The cost of the initiative is listed with the return on investment (ROI) based on number of years. Finally, the risk reduction is listed across from each strategy. For example, establishing a remote manufacturing facility would reduce the risk rating by 30 points for hurricane, winter storm, and tornado for a total risk reduction of 90 points. However, this strategy would cost $5 M, but it would have a very short return on investment (0.018 years).

Table 13.7 Strategies to address vulnerabilities.

Components	Vulnerabilities	Strategies
Economic	Single source of revenue based on production and sale of products; limited backlog of orders that can be filled based on current inventory; insurance policy does not cover lost revenues	Diversify customer base; establish a remote manufacturing facility; establish subcontract or mutual aid agreement with partners to divert work and continue operations utilizing available capacity; modify operations and offer different products based on available resources; establish a captive insurance plan*
Resource accessibility	Workers unable to travel to work when roads are closed; workers vulnerable to contracting disease, potential exposures, caring for sick family	Establish remote work policies, procedures, and IT infrastructure; establish pandemic policies and plans that enable workers to gather at work while preventing and mitigating the spread in a safe manner; require vaccination or negative testing to protect the workforce; establish contract with temp agency to quickly increase staffing when needed
Supply chains	Single source of supply	Identify redundant sources of supply; increase inventory reserves prior to hurricane season
Social and cultural capital	Segment of workers that have comorbidities which could result in more severe illness from the virus	Require negative test results to prevent work-place exposures
Safety and security	Lack of cyber-security on network; workers more likely to become injured during recovery operations, or ill from home-transmission	Adopt NIST cybersecurity standards appropriate for business; conduct safety training prior to hurricane season related to chainsaw, electrical, and downed power line safety.
Infrastructure	Dependency on delivering raw products over roadways; inadequate bandwidth for working from home; network is not secure	Consider alternate delivery methods: train, barge, drone. Improve IT infrastructure to ensure adequate VPN connections and software licenses
Governance	Policies do not address working from home, leave during disasters, implementation of pandemic mitigation and prevention measures	Update HR policies in coordination with business continuity plan
Natural Resources	N/A	Adopt environmental management system (e.g., ISO 14001 standard)

*Captive insurance** is a legally registered insurance entity that enables a business to expense premiums paid to a separate legal entity that it owns for the purpose of covering risks that traditional insurance companies do not cover. This allows the business to retain the funds set aside for claims not used.

Table 13.8 Risk reduction and cost analysis.

	Risk Reduction Associated with Each Initiative						
Strategies	Hurricane R: 75	Pandemic R: 60	Winter Storm R: 40	Tornado R: 38	Cyber Attack R: 80	Total	Cost (ROI (yrs))
Diversify customer base;	5	5	5	5	0	20	$100k (0.25)
establish a remote manufacturing facility;	30	0	30	30	0	90	$5M (0.018)
establish subcontract or mutual aid agreement with partners to divert work and continue operations utilizing available capacity;	60	50	30	30	10	180	$20k (9)
modify operations and offer different products based on available resources;	15	10	15	5	0	45	$150k (0.3)
establish a captive insurance plan	10	10	10	10	50	90	$5k (18)
Establish remote work policies, procedures, and IT infrastructure;	10	50	20	10	0h	90	$5k (18)
establish pandemic policies and plans that enable workers to gather at work while preventing and mitigating the spread in a safe manner;	0	50	0	0	0	50	$5k (10)
require vaccination or negative testing to protect the workforce;	0	50	0	0	0	50	$1k (50)
establish contract with temp agency to quickly increase staffing when needed	20	30	5	10	5	70	$2k (35)
Identify redundant sources of supply;	60	30	15	10	0	115	$10k (11.5)
increase inventory reserves prior to hurricane season	60	0	0	5	0	65	$20k (2.6)
Require negative test results to prevent work-place exposures	0	40	0	0	0	40	$1k (40)

Adopt NIST cybersecurity standards appropriate for business;	5	5	5	5	60	80	$20k (4)
conduct safety training prior to hurricane season related to chainsaw, electrical, and downed power line safety	15	0	0	5	0	20	$2k (10)
Evaluate alternate delivery methods: train, barge,	15	60	10	10	0	35	$10k (3.5)
drone. Improve IT infrastructure to ensure adequate VPN connections and software licenses	10	50	10	10	60	140	$15k (9.3)
Update HR policies in coordination with business continuity plan	5	30	25	15	60	135	$10k (13.5)
Adopt environmental management system (e.g., ISO 14001 standard)	0	0	0	5	0	0	$10k (0.5)

Based on our example, the resilience measures with the greatest impact on risk reduction are the following:

- Establish subcontract with partner to divert orders and continue fulfillment (180).
- Improve IT infrastructure (140).
- Update HR policies in coordination with business continuity (135).
- Identify redundant sources of supply (115).

However, these four measures cost approximately $55k, which may be within the budget for a large organization. This is the total risk-reduction approach.

Another approach is to prioritize based on the return on investment (ROI) which optimizes risk reduction based on cost. While this may not result in the greatest risk reduction, it will ensure good stewardship of resources by ensuring the greatest risk reduction per dollar invested. Below is the re-prioritized list of resilience initiatives based on a $51,000 budget:

- Require negative testing before work.
- Update HR policies in coordination with business continuity plan.
- Identify redundant sources of supply.
- Establish contract with temp agency.
- Establish remote work policies, procedures, and IT infrastructure.
- Establish captive insurance.
- Establish pandemic policies, plans, procedures.
- Conduct safety training.

Keep in mind that both preparedness and resilience are processes rooted in continuous improvement. We live in a dynamic environment with evolving threats and hazards, some of which are increasing in severity. The capabilities within our organization also need to change and match emerging threats and hazards accordingly. Organizations should not only review and update their hazard list and risk assessment annually, but also update and implement new actions to strengthen their systems and enhance resilience.

The COVID-19 pandemic provides a recent case for resilience. The travel, retail, and restaurant industries were devastated by the pandemic. The casualties which include numerous national chain restaurants, rental car companies, hotels, and airlines are still accumulating. However, some restaurants thrived during the pandemic and struggled to keep up with demand as they pivoted to outdoor seating, established/increased drive-through capacity, established online ordering systems, and teamed with delivery companies to provide contactless delivery.

13.7.2 Implement Actions to Increase Resilience

Leaders are responsible for decision-making. While the above approach provides a framework for becoming more resilient, leaders need to ask good questions to ensure that the methodologies are sound, and the research is thorough. Leaders play a critical role before the crisis by establishing strategies, approving plans, and holding their managers accountable for action and results. While the purpose of this book is to develop crisis-ready leaders, it would be short-sighted to only equip leaders for crisis decision-making during the response and recovery phases, although this requires a unique profile of character traits, knowledge, and abilities as outlined in Chapters 10 and 11. It is equally important for leaders to act prior to the crisis by implementing actions that increase resilience and set their organizations on a course for success and resilience in the wake of a disaster.

13.8   Disruptions-Theory of Constraints

The theory of constraints is a business management concept used to optimize an organization's production by eliminating internal or external constraints. Organizations can be controlled by varying three measures: inventory, throughput, and operational expense. Inventory consists of purchased inputs to a process. Operational expense is the cost of processing the inputs into products. Throughput is the rate at which an organization generates cash through sales. The task is to identify constraints throughout the system and eliminate them so that inventory, throughput, and operational expenses are optimized. To apply the theory of constraints in our context of resilience, we need to examine our organization's performance in these three areas based on the disruptions caused by hazards.

Example 1:   A winter storm could disrupt an organization's inventory and cause a constraint in inventory which disrupts the entire system. To become more resilient, the organization could increase their inventory on-hand prior to forecasted incidents to ensure that inventory does not become a constraint during the disaster.

Example 2:   A city could pre-position salt and salt trucks at key points along an interstate prior to the storm to ensure continuity in road clearing. In my business, we prevented weather impacts on our travel by watching the forecast and making the decision to depart for business trips in advance to avoid travel delays during the incident. This ensured continuity of services for customers in other states who were not aware of the weather incident and expected our services.

In addition to inventory and inputs, organizations should examine their operational processes and throughput to determine potential impacts and disruptions from hazards in their hazard profile. To enhance resilience, organizations should develop policies, plans, and procedures to include sources for alternate facilities, equipment, supplies, and personnel in case of disruption. Business continuity planning can help uncover these potential impact areas.

Leaders should ensure that this planning and analysis occurs within their organization so that the organization minimizes the impact from hazards. The vulnerability analysis outlined above provides a framework for identifying constraints. The theory of constraints provides another lens into an organization's operation.

13.9   Whole Community Approach

We have provided numerous examples related to businesses and organizations. Before we depart from that theme, I want to underscore the applicability of the whole community approach for non-governmental entities. Whether a business, church, school, non-profit, or association, these organizations should connect with the public sector and develop relationships with their counterparts before the incident. These partnerships could be as informal as an introductory phone call to let them know you exist and obtain updates on community planning, preparedness, and outreach. As a business owner, I remain connected with economic development organizations, representatives from the building space that we lease, airport authorities, and others, in addition to those more formal contacts with emergency management clients. If you are in the chemical industry, you should be in regular contact with your local fire department, Local Emergency Planning Committee, and regulators. If you operate a school, hopefully you have made frequent contact during COVID with your local health department, but this would be a good idea when COVID is not a current threat as well. The point that I am making is that the private sector should engage with the public sector more regularly to maintain contact and to stay informed regarding community hazards, risks, and preparedness opportunities and initiatives.

During COVID-19, many in the public sector struggled to make the right connections with the business community, understand the impact of COVID, and determine what role the public sector could play in restoring the local economy. Additionally, many elected officials acted out of desperation to pass bills that artificially stabilized the economy through forgivable loans, increases to unemployment payments, and changes in rent agreements, but neglected to address the root causes of the problem thereby perpetuating and exacerbating problems such as supply-chain shortages, lack of labor, uncertainty regarding workplace health and safety, travel restrictions, etc. A whole community approach to addressing the problems of the pandemic could have had a very different and positive outcome with input from community stakeholders.

The public sector's lack of whole community engagement has made the cure worse than the problem in some cases:

- Early in the COVID-19 crisis, the government was overly focused on public health effects from COVID-19 and neglected the economic, educational, and psychological impacts of the restrictions. While these impacts may not be quantified in a comparable manner, leaders are responsible for their decisions and impacts in each area. Studies are now showing the significant increase in suicide ideation, hospitalization, and deaths caused by these decisions. A whole community approach ensures that representatives from each of these domains have input for leaders. Leaders have the difficult task of weighing the impacts and making decisions that balance the benefits and risks in a responsible manner. Ethics analysts will study, evaluate, and report on these decisions with the hope of influencing a proper balance. My point is that early in the crisis, many decision makers lacked representative analysis from these domains from the whole community of stakeholders.
- In an attempted to lighten the burden of renters who have lost their jobs or income, the government prevented landlords from evicted renters for non-payment but still required landlords to pay the mortgage company and bills. Perhaps engagement with this community of stakeholders could have provided relief to landlords in the form of rent payments for those that lost income or jobs. The federal government attempted to correct this by dispersing funds to states but left it up to states to correct the problem.
- By increasing unemployment benefits, the federal and state governments made it more economic for some workers to stay home instead of returning to work when it was safe to do so. This created a shortage of service workers causing many hotels, restaurants, small businesses, and suppliers to close or reduce output significantly. In the case of suppliers, this created a ripple effect throughout the economy with various supply chain shortages.
- International restrictions on shipments and similar workforce constraints abroad showed the extent of dependence on parts and materials from outside one's country. For example, a shortage of computer chips manufactured in China slowed production of cars in the US affecting car manufactures, dealership, consumers, and downstream markets such as availability of rental cars and used cars. This resulted in skyrocketing prices in both markets.
- Lack of swift decisions and implementation of health and safety standards in different jurisdictions contributed to travel delays, disruptions, and confusion. While States control the prevention and control measures within their state, the lack of a centralized clearinghouse disrupted interstate commerce. For example, travelers to New York from designated states were required to quarantine upon arrival and show proof of a negative COVID test within days of travel and upon arrival. While the federal government cannot impose less restrictive standards, the federal

government missed an opportunity to coordinate among the states and facilitate harmonized requirements to improve risk communication and safety early in the crisis.

Learning from the business community and the challenges they faced fostered problem-solving and solutions which both the private and public sector could solve together. In some cases, this resulted in positive changes. Consider the following examples:

- Shuttered restaurants approached elected officials and zoning boards to allow blockading some streets so that they could continue to operate their business outdoors. Some used CARES Act funding to build semi-permanent structures and tents to provide needed ventilation outdoors while providing shelter from the cold and rain.
- Businesses concerned about liability over workplace transmission took the initiative to improve sanitation, physical distancing, and barriers. State and federal regulators heard concerns from employers and developed Emergency Temporary Standards (some of which have become permanent) to provide standards for assessing risks, implementing controls, and educating employees about COVID-19. Regulators listened to employer concerns and sought advice from industry groups such as the AIHA® to inform their rule making.
- Economic development offices pivoted to partnering with businesses to help solve supply chain disruption issues, and provided grants to aid in the implementation of infrastructure changes needed to protect workers and customers.

13.10   Setting Conditions for Successful Outcomes before the Incident

This chapter has elaborated on the concept of resilience and provided a framework for evaluating resilience in different contexts. It should be apparent that a crisis-ready leader is engaged in building a resilient community long before a crisis occurs. By building resilience into the organization, a leader sets the environmental and organizational conditions needed to derive a successful outcome from a crisis. Resilience abates the severity of the crisis impact and provides leaders with the propensity for recovery rather than a disaster spiraling out of control.

As a leader, you will likely face a crisis in your organization or community. If the structure of your system is not resilient, it could break resulting in collapse or a lengthened state of the crisis, neither of which is productive. A leader's role is to advance the vision of the organization by accomplishing the purpose through a strategy. Resilience minimizes the disruption and ensures a quick rebound from stressors on the system. What are your organization's vulnerabilities? What steps can you take today to build resilience into the system?

Summary of Key Points

- Resilience is the capability of a strained body to recover its size and shape after deformation caused especially by compressive stress.
- Resilience indicators are specific to the system but may be organized by various capitals within the system: human, infrastructure, financial, intellectual, and governance.
- The crisis leader's goal is to engage in building resilient systems before the crisis, and to make good decisions early in the crisis to steer the response into recovery while minimizing the impact.
- A systematic approach to increasing resilience ultimately reduces risk.

- The Theory of Constraints model may be helpful in identifying how and where disruptions can occur within a system.
- A whole community approach ensures diversity and inclusivity in the decision-making process. While the crisis may primarily affect one lifeline within a community, the solutions have the potential of causing subsequent problems that may be worse than the original problem.

Keys to Resilience

This chapter focuses on resilience so rather than repeat it, we can summarize the key to resilience as, "engage the whole community of stakeholders in planning, preparing, analyzing, and building resilience before the crisis."

Application

1) How do you define resilience in your context or organization? Who is responsible for building resilience and how do they engage the whole community?
2) In this chapter, we presented a framework for resilience and various types of indicators. This is not the only approach. Describe the framework used in your context or organization.
   a) What are the macro-indicators for determining resilience?
   b) What are the metrics used and how are they weighted to evaluate your level of resilience?
3) Read "Ensuring Supply Chain Resilience: Development and Implementation of an Assessment Tool" and conduct a self-assessment based on the Vulnerability and Capability questions starting on page 71. Reference: https://www.researchgate.net/publication/264598430
   a) What capabilities do you have in place for resilience?
   b) What are your organization's vulnerabilities to disaster?
   c) What gaps have you identified?
4) What role can you play in creating resilience in your context? Provide some examples based on the risks your organization faces.

References

1 National Institute of Standards and Technology (NIST) (2020). Community resilience. https://www.nist.gov/topics/community-resilience (accessed January 8, 2022).

2 University of South Carolina. Hazards and Vulnerability Research Institute index page. https://artsandsciences.sc.edu/geog/hvri/bric (accessed January 9, 2022).

3 Pettit, T., Croxton, K., and Fiksel, J. (2013). Ensuring supply chain resilience: development and implementation of an assessment tool. *Journal of Business Logistics* 34. doi: 10.1111/jbl.12009.

4 U.S. Fire Administration. Wildland Urban Interface. https://www.usfa.fema.gov/wui (accessed January 8, 2022).

5 U.S. Fire Administration. What is the WUI? https://www.usfa.fema.gov/wui/what-is-the-wui.html (accessed January 8, 2022).

6 Zuzak, C., Goodenough, C., Stanton, M. et al. (2021). *National Risk Index Technical Documentation*. Washington, DC: Federal Emergency Management Agency.

7 U.S. Centers for Disease Control Agency for Toxic Substances and Disease Registry. Social vulnerability index. https://www.atsdr.cdc.gov/placeandhealth/svi/index.html (accessed January 8, 2022).

Further Reading

1 Federal Emergency Management Agency (2020). Community resilience indicator analysis. https://www.fema.gov/sites/default/files/2020–11/fema_community-resilience-indicator-analysis.pdf (accessed January 8, 2022).

2 Cutter, S., Ash, K., and Emrich, C. (2014). The geographies of community disaster resilience. *Global Environmental Change* 29: 65–77.

3 University of South Carolina Hazards and Vulnerability Research Institute Publications Page. https://www.sc.edu/study/colleges_schools/artsandsciences/centers_and_institutes/hvri/publications/index.php (accessed May 28, 2022).

14

Navigating from Crisis to Recovery

Hopefully, Chapter 13 has inspired you to pursue building resilience in your organization and within your context. Inevitably, external influences will cause a crisis. A prepared leader will not make decisions that worsen the crisis into a disaster; rather a prepared leader will make decisions that navigate through the crisis on the swiftest course to recovery and return to normalcy. Operating within a resilient context will facilitate that return to normalcy with the possibility of even rebounding into a better state.

This book contains various aspects of what it takes to become a crisis-ready leader. We will conclude by reviewing the pathway covered throughout the book and compiling the various elements into a formulation for successfully navigating from crisis to recovery.

Lesson Objectives

After reading this chapter, you will be able to:

- Summarize the role of a crisis-ready leader before, during, and after the crisis.
- Develop a plan of action to ensure your organization and leadership are ready to navigate the next crisis.

14.1 Summary of a Leader's Role in Crisis Leadership

Throughout this book, we have covered a series of topics that are relevant to a leader and their organization before and during the crisis. Each chapter presented a different building block. Table 14.1 and the remainder of this section summarize the relevance of each chapter and how these chapters relate to each other.

A crisis-ready leader:

- Establishes vision, achieves the mission, and holds steadfast to uncompromising values.
- Influences internal and external stakeholders to adopt the vision and commit to direction.
- Prepares for the unexpected by building capabilities based on risk.
- Allocates resources based on greatest risk reduction.
- Adopts and implements an incident management system.

Crisis-ready Leadership: Building Resilient Organizations and Communities, First Edition. Bob Campbell, PE.
© 2023 John Wiley & Sons, Inc. Published 2023 by John Wiley & Sons, Inc.

- Focuses on the long-term and therefore invests in building resilience in their organization and community; during a crisis, leaders leverage any additional resources to achieve resilient recovery outcomes.
- Commits to self and organizational continuous improvement; recognizes and pursues the character traits, knowledge, skills, and abilities of a crisis leader.
- Understands how stress can affect their judgment and has adopted measures to counteract the influence of stress.
- Monitors and measures performance to achieve positive outcomes; adjusts direction when warranted to abandon ineffective courses of action for more effective outcomes.

Table 14.1 Crisis leader roles before and during a crisis.

Pre-crisis	During Crisis
Identify threats and hazards	Enhance situational awareness
Define your operating context	Implement critical thinking for accurate problem definition, COA development, and solutions
Conduct stakeholder identification and analysis	Utilize system to make progress toward stabilization and restoration.
Define the organization's vulnerabilities and challenges	Manage stress
Establish an incident management system	Implement the IMS
Develop attributes of a crisis leader	Self-reflect on performance
Be prepared	Identify opportunities to mature
Build resilience	Develop recovery outcomes that build resilience

14.2 Pre-incident

During the pre-incident phase, leaders should identify the applicable threats and hazards that create risk for their organization. While there are some excellent starting points for considering these risks, leaders should also identify emerging trends and conditions which could result in novel risks to their organization.

Leaders should not only understand their internal operating context such as their plans, organizational roles and responsibilities, but they also need to understand their external context within their community and how their organization interfaces with external entities. This includes identifying their external stakeholders such as community-elected officials, government contacts in emergency management, public health, economic development, emergency services (fire, police, EMS), utilities, community groups that could affect operations, volunteer organizations, general population and their respective demographics, and trade associations. When facing a crisis, its best to know your resources, contacts, incident management system, and lines of communication.

Leaders should engage in risk-based planning with their organization to understand and validate assumptions, verify roles, responsibilities, and authorities, and validate the relevance of the plan for the applicable threats and hazards. Due to limited resources and to ensure proportionality in the allocation of resources to build capabilities, leaders should become familiar with the risk associated with each threat and hazard. Combined with the organization's capabilities, a gap analysis will present vulnerabilities so that the leader can determine risk acceptability based on risk tolerance and resources available to improve the risk profile. Leaders are responsible for managing and binding risk. Unfortunately, few organizations have the resources to reduce all risks to negligible levels.

Each organization should have an incident management system. This system may be comprised of a command and coordination system to organize unity of effort (such as ICS and EOC), communications and information management systems (such as a live-updated GIS-enabled application), and resource management. Establishing these components, training your team on working within these systems, and ensuring the effectiveness of this system facilitates the leader's incident management during the crisis. Once the crisis occurs, the leader can establish and communicate their intent and objectives throughout the organization and ensure unity of effort. Resources can be assigned and tracked for accountability and impact on schedule. Stakeholders are accessible and know their role. Progress toward accomplishing objectives can be tracked by all so that those that are dependent on another objective have the situational awareness needed to coordinate. Lifelines can be stabilized and restored. Good systems enhance situational awareness throughout the organization enabling mid-tier leaders to stay informed, collaborate, and develop relevant courses of action for leaders.

As we uncovered in Chapter 10, Profiles in Crisis Leadership and Chapter 11, Attributes of a Crisis Leader, a crisis-ready leader requires a unique set of attributes, knowledge, skills, and abilities beyond traditional steady-state leaders. Crisis leaders should build a personal and professional development plan that utilizes the self-assessment tools outlined in these chapters to learn more about themselves and their team members. Developing character attributes requires intentionality and personal accountability. This is not a short-term endeavor, but rather a life-long learning process. Application of the requisite skills can be tested during realistic, scenario-based training, and exercises.

Being prepared is a team effort; it's a culture. Developing this culture requires an ongoing, trusted system. The preparedness cycle fosters continuous improvement. The crisis-ready leader engages in the strategic planning process to ensure that their vision is articulated so that goals can be established to progress toward the vision. A crisis-ready leader monitors progress, rewards milestone achievements, and holds their team accountable to stay on course. This occurs through advocating for the resources needed to achieve their vision and goals, empowering their people through professional development, and setting performance and behavior expectations. Taking bold actions often involves external parties. A leader should leverage their relationships within the community to facilitate the work of their organization. Under the right conditions, your team will work through the mechanics of the preparedness cycle (i.e., assessing risks, planning, organizing, building, training, exercising, and improving). A good leader creates momentum, encourages their team to continue in the direction of progress, and greases the bearings of the organization to accelerate activity and ultimately preparedness.

But crisis is inevitable. Being prepared to respond and recover is not enough. Building a resilient organization is also a long-term process that begins with standards, codes, and designs in the built environment. As extreme weather events and climate change impact the frequency and severity of incidents, upgrades and enhancements to the built environment are necessary. Additionally, organization plans must strengthen an organization's essential functions and capabilities with flexible options, alternatives, and redundant systems. Stress testing organizational capabilities can provide some insight into an organization's level of resilience. Organizational resources such as human, supply, utilities, and information must also be strengthened and tested to ensure adequate access to these essential resources during a crisis. A resiliency assessment can reveal the strength and vulnerabilities of the system. The results can be integrated into the organization's planning and budgeting to ensure adequate resource allocation based on risk and vulnerability.

With these capabilities and systems in place prior to an incident, leaders will be ready to handle the next crisis.

14.3   During the Crisis

When an incident occurs, leaders should immediately assess the rate of change in their risk profile to stay ahead of the incident. The crisis requires the leadership team to maintain good situational awareness which requires reliable, timely, and accurate sources of information that can be integrated, displayed, and compiled for processing. Having a robust incident management system in place prior to the incident will enable the swift receipt and processing of new information so that leaders can develop courses of action and improve their leader's decision quality.

As new data and information are received, a leader should foster their team's collaboration to ensure critical thinking can be applied. This will not happen by accident. Putting the right people with critical thinking and analytical skills in leadership positions enables this process to be effective. A good leader will ask good questions to help define the problem. In complex adaptive systems, there are likely multiple problems interacting which can cloud judgment and redirect focus away from the real problems. A leader must not stop at problem definition. In order to solve the problem, the root causes and contributing factors must be determined. A team of experts can be helpful in working through the technical details. Knowing who to appoint to these roles before the incident will ensure that a reliable and dependable panel of experts will develop relevant courses of action that address problems and their root causes. Again, this will improve decision by having multiple good options available. The leader is ready to make decisions and allocate the resources to move the response and recovery forward. While every decision may not be as effective, a leader must continue to charge forward toward stabilization and restoration through good decision-making.

A leader must also establish their intent and objectives so that resources can be directed under unity of effort. This is how information, intelligence, resources, and innovation sharing happens – through unity of effort. Leaders must work with their team to define pathways to stabilization and restoration through intermediate objectives, requesting and allocating resources, monitoring the effectiveness of their activities, and making adjustments to stay on course. Effective pre-incident plans clearly define the pathway to stabilizing lifelines so that recovery and restoration can continue. Working through multiple operational cycles, leaders should ensure consistency and unity with other leaders. With a sound process and good leadership, the rate limiting factor to recovery will be availability and cost of resources.

Leaders must proactively manage the stress induced by the crisis. First, leaders must be aware of how stress effects their mind, body, and emotions in order to develop strategies to overcome the effects of stress. Second, leaders must engage in activities to maintain their mental health and wellness such as exercising and journaling. Third, leaders must participate in stress-induced, scenario-based training and exercises to build the muscle memory necessary to be successful under the pressure of a crisis. Finally, leaders must develop decision-support tools such as checklists, plans, procedures, and techniques to stay focused in driving the incident to a conclusion and into recovery. If a leader is prepared for stress and recognizes their limits, then the leader will be capable of making sound decisions while under pressure.

Leaders must implement their incident management system. This includes assigning personnel to positions within the ICS, but also implementing information systems that enhance situational awareness, managing resources, and establishing communication systems both internally and with external stakeholders.

As a leader navigates the crisis, there will undoubtedly be moments of truth that determine whether the leader was prepared or unprepared. Leaders must avoid political spin and embrace the facts of the situation while holding onto the vision they have for the future. This means confronting shortcoming through self-reflection and determining how they can improve their leadership for future crisis situations. All of our profiled crisis leaders from Chapter 10 highlighted humility as a key character trait that enabled them to admit when they were wrong and move forward in a spirit of improvement.

During the pre-incident phase, leaders prepared for the highest risks. During a crisis, that preparedness will reveal the organization's capabilities and opportunities to mature. While initial reflections on an incident will lead to biases of preparing for the last incident, leaders need to put the last incident into the context of risk from all-hazards. Risk-based planning will lead to enhanced capabilities that cross-cut all hazards.

As the incident unfolds, leaders should be developing recovery outcomes that build resilience for the long-term and future incidents.

Each of these aspects of a leader's role during a crisis will determine the leader's ability to navigate the crisis into recovery and beyond. But as it is outlined, much of the leader's success is determined before the crisis.

14.4 Post-crisis

As a crisis is resolved and normalcy is within reach, a leader should self-reflect on their leadership and the team's performance. When a leader truly learns from the past whether history, experience, or others, then the leader is ready to rebuild and improve. A good hotwash led by a trained facilitator can yield valuable feedback. There are multiple frameworks for conducting a gap analysis or hotwash, but I prefer the DOTMLPF framework used by the military. This covers several specific areas which can be more beneficial to leaders than other frameworks if used properly. In my experience conducting hotwashes after exercises or during vulnerability assessments, few people are inclined to identify "leadership" issues or gaps. This is where a leader with a healthy relationship with their team and genuine humility can invite critique and suggestions for improvement. The DOTMLPF framework covers doctrine/plans/procedures, organization, training, material/equipment, leadership, personnel, and facilities/infrastructure. Problems can stem from several of these areas and contribute to a failure in execution.

Best Practice: Hotwashes and after-action reporting require several different techniques and systems to gather honest feedback. After we conduct an exercise, we immediately discuss how it went from the participant's perspective. For example, we ask them to identify three things that went well and three things that could have gone better. After settling down, we hand out feedback forms to get written feedback from individual perspectives; these may include responders, observers, role players, and mock casualties. Then, our evaluators compare notes and discuss any discrepancies in observations. Using these various techniques and compiling perspectives, we are able to develop an accurate and reliable after-action report.

Problem identification is insufficient. We should strive to determine the root cause and contributing factors through interviews and analysis. This informs recommendations to ensure that both the problem and the root cause are addressed. Two techniques that we have used successfully are the five Whys and "How did we get here?" The five whys technique involves asking why repeatedly to "peal back the layers of the onion." Each time asking, "why did that happen?" The benefit is that you may be able to identify the root cause(s), but sometimes the team may be reluctant to callout a person which could result in blame. The second technique uses a team-based approach where the leader asks, "how did we [emphasis added] get here?" This approach avoids blame and assumes responsibility at the team level. This is a technique used in aircraft and industrial accidents to focus on the issues, and avoid the politics of who is to blame.

Improvement plans identify recommendations and prioritize them for resourcing and implementation. Be cautious to avoid limiting the improvement based on the last crisis. Instead, consider the potential of different circumstances and the need for solutions which can be applied in multiple situations.

This summary of a leader's role before, during, and after the crisis should help frame your plan of action for becoming a crisis-ready leader.

14.5 Developing a Plan of Action

At this point it is time to determine your plan of action. What steps do you need to take to advance your leadership from a traditional leader to a crisis-ready leader? You may want to review your responses from previous chapters, especially Chapters 10 and 11. Use the following exercise to guide you through the process.

14.5.1 Organizational Readiness

1) Has your organization identified hazards and assessed risks to the organization or community? Have you reviewed and verified that this assessment addresses emerging hazards to accurately estimate your organization's risk profile? FEMA's National Risk Index may be a helpful tool to complete this assessment.
2) Has your organization developed plans such as pre-incident, emergency response, recovery, mitigation, and continuity of operations plans? Have you reviewed the assumptions, applicability, relevance, roles, and responsibilities to ensure that the plan is adequate for the risks identified above?
3) Has your organization adopted and implemented an incident management system? Have roles been assigned to the right people/positions and has this been tested in an exercise or real-world incident?
4) Has your organization identified its internal and external resources, capabilities, and capacities? Have you met with external stakeholders to coordinate your organization's plans and IMS? Do you know how to access the external resources you may need? How long will it take to receive those resources?
5) Have you participated in the strategic planning process to cast vision, establish goals, and review initiatives to ensure they align with your priorities for preparedness and resiliency?
6) Does your organization provide the training needed for key personnel to fulfill their emergency response roles? Are your personnel equipped to perform the duties that may be required during an emergency?
7) Has your organization conducted realistic exercises to test, verify, and validate your capabilities?
8) Has your organization determined your vulnerabilities?
9) Have you considered disaster-related risks in your organization's risk profile? Have you managed risks equitably by allocating resources to address all vulnerabilities based on risk profile?
10) Have you identified resilience initiatives to prevent and mitigate the impact from a crisis?

14.5.2 Leadership Readiness

1) Based on the application exercises in Chapter 11, what character traits, knowledge, skills, and abilities do you need to work on? Have you identified a personal and professional development strategy with milestones, schedule, resources needed, and accountability to pursue this development? If not, list out the areas of personal and professional growth that you identified during these exercises. Research opportunities to work on gaining knowledge and developing skills and abilities.

2) Have you identified a mentor or advisor to hold you accountable for your growth plan? Identify and discuss with mentors or colleagues your desire to improve on character traits and invite them to work with you through accountability and feedback. How often are you listening, tracking, and implementing their advice? Keep a journal of their feedback and document which inputs you have adopted and implemented, and which inputs you have dismissed.

3) Have you trained under or experienced stressful conditions to reveal your vulnerability to stress-induced effects on judgment? If not, find an opportunity to challenge yourself whether in an exercise or simulation that taxes your mental and physical resources. It is important to recognize your limits and how your judgment can be compromised under stress. What do you need to do to improve your mental health and wellness so that you are ready for a crisis? What planning, training, exercises, and personal disciplines are needed to enhance your readiness for a stressful crisis?

4) What resources are available to maximize your situational awareness? What is missing? Are there additional tools or information sources needed to enhance situational awareness?

5) What kind of decisions do you anticipate making at the onset of a crisis? What resources do you rely on to provide you with reasonable courses of action to ensure you are making the best decision?

6) What role do you fill in your organization's command and coordination system (i.e., ICS, EOC, etc.)? Who do you report to? Who reports to you? What external entities or people will you likely need to work with? Have you participated in advanced planning with them?

7) Have you participated in realistic, scenario-based exercises that test your capabilities and capacity? What are you learning about yourself and your organization that need improvement? What kind of feedback are you receiving from others on your leadership? Is this consistent with your self-assessment? What are you planning to do to improve?

8) How familiar are you with your incident management system components?

9) What indicators will you use to determine stabilization, restoration, and progress toward normalcy?

14.6 Conclusion

Your responses to the questions listed above should provide some indications of strengths and weaknesses in your ability to act as a crisis-ready leader. It should also provide you with a roadmap of opportunities to improve your leadership and your organization. The next step is to act on these items. Set personal and professional goals to enhance your leadership and engage in organizational readiness.

Throughout this book, we have listed additional resources to go deeper into various topics. Over the last five years, I have engaged in further developing training materials and this body of knowledge related to crisis leadership, situational awareness, decision-making, and various applications to public-health emergencies, hazardous materials response, and economic recovery as my company has designed and developed training in these topic areas. Through the instructional system design process, we have met with various stakeholders and SMEs in the private and public sector in all levels of government to determine their training needs and properly design training to address these needs. This process continues to inform our body of knowledge, case studies, and training development. Because of the demand and ongoing improvements, we have established a website with additional resources to include recorded videos that cover several of the topics outlined in this book. We invite you to stay current and utilize these resources at www.asg-inc.org/crisis-ready-leader.

Index

Note: Page numbers followed by "*f*" and "*t*" indicate figures and tables, respectively.

a

active threat, application   90–91
all-hazards incidents   10, 149
Al Qaeda in Iraq (AQI)   139–141
application-based training   126, 127, 182
asset poverty   136

b

betacoronavirus   1
Brock Long   134–139

c

capabilities-based planning   40, 41
captive insurance   200
cascading impacts   197
catastrophe   14
catastrophic disaster   20
Chad Hawkins   149
    COVID-19 Response Incident Management
        Team Incident Commander   149–152
chemical, biological, radiological, and nuclear
        (CBRN) weapons   16
combined public–private operating context
        28
    COVID-19 (Global Incident of Significance)
        29–30
    El Dorado Chemical facility, Bryan   28
    Superstorm Sandy Disaster   29
    West Fertilizer Company   29
communication systems   11, 15, 183
community lifelines   93–106, 137

construct   95–96
detecting and reporting unstable
        components   103
incident priorities, establishing   99
lifeline assessment and status   98–99
lifeline conditions and status
        reassess   101–102
lines of effort, stabilizing   103–104
logistics and resource requirements   100–101
operationalizing lifelines   100
organizational lifelines and components
        adapting   103
organizational perspective   102–104
organizing response activities   100
stabilization and recovery   102
community lifelines toolkit
    benefits of   96–97
    limitations with   97–98
competency   127–128
component impact analysis   198
components   94, 95*f*
Continuity of Operations Plans (COOP)
        103, 104
COVID-19 pandemic   1, 13, 85, 86, 112, 149, 206
    Global Incident of Significance   29–30
    mass vaccination, application   91
crisis/crises   12–14, 17, 20, 43, 45, 47, 75, 112,
        114, 115, 128, 145, 148, 207, 211, 213–215
    assign resources   46
    characteristics   13*t*

communication and   47
conditions for success, creating   43–45
coordinate with stakeholders   46
goal and incident objectives
    communicate   45–46
goal and objectives, establish   45
leading through   45–47
progress and control outcomes
    monitoring   46
stress   109, 110, 115, 129
crisis leaders   17, 37, 77, 93, 124, 128, 133, 145, 157, 158, 160, 169, 170, 177, 213
additional advice   168–169
attributes   157–175
crisis leadership skills, developing   170
essential character traits   157–161
knowledge   161–165
skills   165–168
crisis leadership   10, 37, 131, 134, 136, 138, 140, 142, 144, 146, 148, 150, 152, 154, 156, 157
leader's role in   211, 212*t*
profiles in   133–156
crisis management   13, 20, 30, 88, 167, 179
strategies   13*t*
crisis-ready leaders   47, 131, 157, 164, 177, 204, 207, 211–213, 216, 217
crisis-ready leadership   9, 23, 37, 49, 59, 71, 93, 109, 119, 179, 189
critical thinking process   71–77, 79, 87, 88, 120–122, 214
comprehension   73–75
courses of action   77–78
perception level   72–73
problem-solving   76–77
projecting or predicting future   75–76
Cuban Missile crisis   13
culture of preparedness   14, 42, 44, 48, 134–136, 139, 193

d

decision makers   1, 59, 64, 65, 67, 72, 74, 77, 80, 81, 84, 85, 87, 98, 115, 122–124, 128
decision-making process   12, 46, 50, 57, 65–68, 71–73, 75–89, 91, 110, 114, 116, 119, 137, 181
data quality objective model   83–85
implementing decision   86–87

making good decisions   82
resource constraints   79
subjective, empirical model   85–86
time constraints   79–82
decision quality   57, 65, 71, 78–81, 163
decision-support tools   119, 125–126, 128, 163, 214
decision theory   57, 71, 72, 74, 76, 84, 86–88, 93
Derrick Vick   145
President of Freedom Industries   145–149
disasters   12, 14, 17, 20, 36, 39, 40, 42, 44, 45, 95, 112, 134, 135, 137, 138, 162, 177
disruptions   96, 97, 99, 103, 189, 190, 196, 197, 200, 205–208
disruptions-theory of constraints   205
distress   112, 113, 115, 116, 119
duration   50

e

economic indicators   17
El Dorado Chemical facility, Bryan   28
emergency communications center   29, 150
emergency management   30, 53, 121, 125, 135, 169, 177, 181, 189, 212
emergency response plan (ERP)   16, 26, 27, 35, 36, 68
emergency support function (ESF)   96–98, 137, 150
environmental factors   66, 68, 69, 71
essential character traits   157–161
humility   158
integrity   158–159
moral courage   160–161
people smarts   159–160
eustress   112
exercise design   68
expertise   127–128
external factors   57, 59, 65, 66, 68

f

Federal Emergency Management Agency
    (FEMA)   2, 9, 29–31, 93, 96, 110, 121, 134–139, 162, 168
community lifelines   93–106
financial capital   191
financial crisis   145, 147
financial resources   15, 30, 43, 103, 181, 182

Frank Patterson 142
  West Fertilizer Company Incident response
    142–145
funding systems 41–42

g
geographic information systems 72, 88, 121
geopolitical unrest or conflict 16
good leadership 17, 214
governance/systems 191

h
Hazard Identification and Risk Assessment
    (HIRA) 49–50, 55, 180, 181, 193
hazardous material (HazMat) 12, 15, 16, 73, 74,
    142, 182
  incidents 25
hazards 9–10, 15, 16, 37–39, 41, 47–52, 54, 73,
    75, 193, 194, 200
  expanding scope of 38
  increasing severity of 38–39
health 119–120
Hazardous Materials Environmental Response
    (HMR) 27
Hot Wash 21
human capital 191

i
incident action plan 45, 67, 99, 100, 144, 149,
    150, 163, 181
incident command system (ICS) 18, 23, 24, 24*f*,
    25, 29, 30, 35, 36, 66, 72, 123, 127
incident management systems 9, 53, 66, 79, 105,
    150, 211–214, 216
incident management team 31, 134, 149, 163
incident objectives 24, 45, 46, 93, 97–101, 143,
    144, 150
incidents 10, 11–14
  catastrophe 14
  crisis 12–14
  disaster 14
  emergency 11–12
  spectrum of 10
information overload 72, 87, 102, 123–125, 137
information systems 17, 46, 72, 182, 191, 199
infrastructure 191

insurance 135, 191, 196
integrity 74, 145, 158, 159, 167
intellectual capital 191
international news 16
inverted-U arousal-performance model 111*f*

j
judgment effect 109
justice and equity 17

k
knowledge 161–165
  multidisciplinary knowledge 164–165
  plans, organizations 162–163
  self-awareness 161–162
  situational awareness 163–164
knowledgeable personnel 79, 80
knowledge-based training 126, 127, 129, 182
knowledge gap 79, 161

l
labor supply shortage 1
large scale disaster 31–32
leadership 53, 140, 143, 146, 168, 181, 183, 185,
    187, 211, 214–217
  attributes 157, 170
  skills 145, 152
Lieutenant General (Ret) H.R.
    McMaster 139–141
  counterinsurgency against Al Qaeda,
    Iraq 139–141
life-and-death decisions 80, 160
lifelines 94–102, 104–106, 137–139
likelihood 16, 50–52, 54, 77, 122, 123, 180, 194,
    200
limiting factor 94
lines of effort 94

m
Major General (Ret) Dana Pittard 152
  leading the campaign against ISIS in
    Iraq 152–155
mobile hazards 26, 27
moral courage 160, 161
multidisciplined response 32
  consensus standards, operations 33

credentialing and certifications   32
HazMat detection   33
HazMat response training and
    certification   32–33
public communication   33
recovery contractors   34
technical specialists   33–34
triage, decontamination, and sampling   33
multijurisdictional response
challenges of   31
dominant forces in   32*f*
escalating scale of incidents   31
multinational context   31–32
multitasking, human brain   123

n

National Incident Management System
    (NIMS)   23, 25
national preparedness, defined   179
natural disasters   12, 15, 16, 20, 36, 50, 52, 54,
    158, 167, 177
natural hazards   38, 191, 193, 194
non-profit organizations   3
norms, rethinking   42–43

o

operations plans   68, 103, 105, 121, 138,
    192, 216
operations tempo (ops tempo)   153, 168, 169
organizational approach   42
organization's capabilities   129, 185, 212, 215
organization's vulnerabilities   207, 208

p

physical fitness   120
plan effectiveness   121
planning   40–41, 50, 68, 102, 120–123, 129, 162,
    163, 181, 185, 186
capabilities *versus* risk-based planning   41*t*
plan of action   216–217
leadership readiness   217
organizational readiness   216
post-crisis   215–216
post-incident phase   10, 15
pre-incident phase   10, 12, 17, 32, 68, 79, 88, 98,
    99, 212, 215
leaders   212–213

pre-incident planning   11, 17, 23, 28, 29, 34, 35,
    45, 46, 66, 123
preparedness   42–45, 47, 48, 134–136, 138, 139,
    179–187, 193
defined   179
national preparedness   179
preparedness cycle   179–180, 180*f*, 181, 183,
    185–187, 193, 213
exercise/testing   183–184
implementation   185–186
improvement   184–185
organizing and equipping   181–182
planning process   181
risk assessment   180
training   182–183
private organizations   25, 28, 96, 97, 102, 165
private sector context   25–26
mobile hazards   27
stationary hazards   26–27
proficiency   127–128
psychological effects   112–114, 116, 117
public–private operating context, combined. *see*
    combined public-private operating context
public responders   26, 27, 34
public safety context   23–25

r

resilience
6 R's of   190
building   189–208
community capacity   191
components of system, hazards
    impact   194–196
conditions for successful outcomes   207
definition   189
economic/financial   191
human well-being/cultural/social   191
implement actions to increase   204
indicators   190–192
infrastructure/built environment/
    housing   191
institutional/governance   191
interdependent components and extent of
    impact   196–197
local, relevant hazards and assess
    risk   193–194
model   190

strategies to strengthen components 200–205
system, definition 193
systematic approach to 193–197
system-wide resilience 200–204
traditional emergency management
    model 192
whole community approach 205–207
resilience indicators 190, 191, 207
resource constraints 79, 84, 127
Response Management Communication Center
    (RMCC) 27
response operations 14, 15, 26, 32, 96, 97, 100,
    101, 143, 144, 152
response organizations 26, 88, 152, 165
responsibilities 24, 27, 32–35, 44, 121, 122,
    181–184, 212, 216
risk 50
risk-based planning 17, 40, 41, 47, 49, 50, 52, 54,
    122, 212, 215
    Hazard Identification and Risk Assessment
        (HIRA) 49–50
    risk calculation 50–52
risk calculation 50–52
    consequence rating 52
    likelihood rating 52
risk profile, civil unrest 53*f*
roles and responsibilities 122
Rumsfeld's Known Knowns model 161

s
safe and secure tomorrow 177
scenario-based training 126, 127, 129, 182, 183,
    213, 214
senior leaders 101, 103, 105, 111, 153
shortfall 94
situational awareness 12, 46, 57, 59, 60, 65, 68,
    69, 71, 72, 75, 79, 81, 87, 99, 102, 104, 151,
    163, 164
    defined 59
situational awareness model 60
    application of 67–68
    comprehension 61–63
    decision-making, relationship 65–67
    individual factors 67
    intervention 64
    perception 60–61
    projection 63–64

task and environmental factors 66–67
skills 165–168
    communication 165–166
    complex problem-solving 166–167
    decisiveness 168
    design thinking 166–167
    trust building 167
sound judgment 107
spectrum of incidents 10, 28*t*
    incident phase 11–14
    organizational context 23
    post-incident 14–15
    pre-incident phase 10–11
stabilization 95
standardized training 68
strategic empathy 158
strategic plan 135, 136, 138, 162, 166, 181
stress 107, 109–117, 119–122, 125, 128, 129, 142,
    148, 214
    definitions 109–111
    effects of 66, 111, 116, 119, 120, 163, 214
    impact of 107, 109, 116, 117
    on judgment effects 114–115
    overcoming to optimize
        performance 119–129
    physiological effects 113–114
    psychological effects 112–113
    spectrum of 109–111
stressful crisis 115, 217
subcomponents 94, 103–105
Superstorm Sandy Disaster 29
supply chain impact 197*t*, 198*t*
suppression systems 66, 123

t
technological hazards 13, 15, 16, 36, 50
technology advances 16
threats 9, 15
    expanding scope of 38
    increasing severity of 38–39
time constraints 71, 79–81, 87
train collision, application 88–90
training
    and exercises 126–127
transportation systems 39, 42, 123
trust 32, 46, 121, 122, 139–141, 144, 145, 152–
    154, 158, 159, 167

u

unified command   34

v

vulnerabilities   37, 39, 45, 47, 48, 50, 53, 54, 193,
        194, 199, 200, 213, 216, 217
vulnerability   39
vulnerability assessment   53–54
vulnerable populations   39–40

w

warning time   50
wellness   119–120
West Fertilizer Company   29
whole community preparedness   43
Wildland Urban Interface (WUI)   192
workers   26, 110, 123, 146, 197, 206, 207
workload   66
workplace hazard   116

Printed and bound by CPI Group (UK) Ltd, Croydon, CR0 4YY

19/07/2023

03237994-0001